177 Topics in Current Chemistry

Springer

Berlin
Heidelberg
New York
Barcelona
Budapest
Hong Kong
London
Milan
Paris
Santa Clara
Singapore
Tokyo

Electron Transfer II

Editor: J. Mattay

With contributions by
R. Bassi, T. Clark, A. Heidbreder,
S. Hintz, R. C. Jennings, J. M. Kelly,
A. Kirsch-De Mesmaeker, J.-P. Lecomte,
J. Mattay, S. Steenken, G. Zucchelli

With 19 Figures, 4 Tables and 111 Schemes

Springer

This series presents critical reviews of the present position and future trends in modern chemical research. It is addressed to all research and industrial chemists who wish to keep abreast of advances in their subject.

As a rule, contributions are specially commissioned. The editors and publishers will, however, always be pleased to receive suggestions and supplementary information. Papers are accepted for "Topics in Current Chemistry" in English.

ISBN 3-540-60110-4 Springer-Verlag Berlin Heidelberg New York

Library of Congress Catalog Card Number 74-644622

Printed in Germany

Typesetting: Macmillan India Ltd., Bangalore-25
SPIN: 10495087 51/3020 - 5 4 3 2 1 0 - Printed on acid-free paper

Guest Editor

Prof. Dr. *Jochen Mattay*
Institut für Organische Chemie
der Universität Kiel
Olsenhausenstraße 40
D-24098 Kiel

Editorial Board

Attention
all "Topics in Current Chemistry" readers:

A file with the complete volume indexes Vols.22 (1972) through 177 (1996) in delimited ASCII format is available for downloading at no charge from the Springer EARN mailbox. Delimited ASCII format can be imported into most databanks.

The file has been compressed using the popular shareware program "PKZIP" (Trademark of PKware Inc., PKZIP is available from most BBS and shareware distributors).

This file is distributed without any expressed or implied warranty.

To receive this file send an e-mail message to:
SVSERV@VAX.NTP. SPRINGER.DE
The message must be:"GET/CHEMISTRY/TCC_CONT.ZIP".

SVSERV is an automatic data distribution system. It responds to your message. The following commands are available:

HELP	returns a detailed instruction set for the use of SVSERV
DIR (name)	returns a list of files available in the directory "name",
INDEX (name)	same as "DIR",
CD <name>	changes to directory "name",
SEND <filename>	invokes a message with the file "filename",
GET <filename>	same as "SEND".

For more information send a message to:
INTERNET:STUMPE@SPINT. COMPUSERVE.COM

Preface

Like the 1994-issue of the series "Electron Transfer", the second volume again covers various aspects of this fundamental process. The articles are concerned with the experimental and theoretical aspects of electron transfer in chemistry and biology. In the latter, emphasis is given to energy transfer, which is also part of photosynthesis.

The concept of electron-transfer catalysis by metal ions is thoroughly discussed in the first chapter on the basis of ab initio calculations, indicating its importance for a variety of fundamental organic reactions. The second article is concerned with transition metals such as Ru(II), Rh(III) and Co(III) which bind to DNA. Upon photolysis, electron transfer processes are initiated, leading in general to damage of the DNA molecule. The use of metal complexes as probes for studying the structure of nucleic acids and potential photo-therapeutic applications are discussed. The third chapter is devoted to cyclization reactions of radical ions, independent of their generation by chemical, electrochemical or photochemical methods. The scope of this new synthetic method has by far not reached its limitations and future ivestigations will prove its potential compared to the radical and ionic counterparts. The role of electron transfer in redox reations between radicals and organic molecules is discussed in the fourth article. In general, electron transfer occurs after heterolysis of an "addition product" (inner sphere). Addition takes place even in these cases electron transfer is strongly exothermic. These findings indicate that inner-sphere electron transfer processes are of more fundamental importance than the outher-sphere reactions of organic chemistry. The final chapter is less concerned with electron transfer than with energy transfer processes. However, the light-harvesting antenna systems of higher plants are as important as the reaction centres themselves, where charge separation takes place. Without these photosystems the world would not exist, at least in its familiar form.

As stated earlier, the present reviews again reflect the interdisciplinary character of electron-transfer research. I am very grateful to the authors for their efforts and in some cases for their patience during the period from submission to printing. The support given while preparing this volume by Dr. Hertel, Dr. Stumpe and their team at the Springer-Verlag is also gratefully acknowledged.

Münster, September 1995 Jochen Mattay

Table of Contents

Table of Contents of Volume 169

Electron Transfer I

Ab Initio Calculations on Electron-Transfer Catalysis by Metal Ions

Timothy Clark

Computer-Chemie-Centrums des Instituts für Organische Chemie der Friederich-Alexander-Universität Erlangen-Nürnberg, Nägelsbachstraße 25, D-91052 Erlangen, Germany

Table of Contents

Topics in Current Chemistry, Vol. 177

The use of ab initio molecular orbital theory to treat electron-transfer catalysis by metal ions and closely related subjects is described. The theoretical principles involved in "hole-catalysis" (acceleration of a reaction by one-electron oxidation) are first examined using the norbornadiene/quadricyclane radical cation rearrangement as an example. The theoretical techniques necessary to obtain reliable results for radical and radical ion systems are also discussed. Examples of calculational studies on hole-catalyzed cycloadditions, sigmatropic rearrangements and electrocyclic reactions are given. The basic principles governing the energetics of electron-transfer between metal ions and organic substrates are described. Finally, calculational examples of electron-transfer catalysis by metal ions are treated. The examples include 1,3-hydrogen shifts, cyclopropane ring-opening, ethylene dimerization, C–C bond activation, and cycloalkane and oxirane ring-opening.

1 Introduction

The concepts of electron-transfer catalysis and so-called "hole-catalysis" [1] are closely related. It is now generally accepted that many organic reactions that are slow for the neutral reaction system proceed very much more easily in the radical cation. Although "hole-catalysis" is now well documented experimentally [2], there is surprisingly little mention of the corresponding reductive process, in which a reaction is accelerated by addition of an electron to the reacting system. Although the concept of "electron-catalysis" is not as well known as hole-catalysis, there are experimental examples of electrocyclic reactions that proceed rapidly in the radical anion, but slowly or not at all in the neutral system [3]. For reasons that will be outlined below, we can expect that, in many cases, difficult or forbidden closed-shell reactions will be very much easier if an unpaired electron is introduced into the system by one-electron oxidation or reduction. Thus, if a neutral reaction **A** → **B** proceeds slowly or not at all, the radical cation ($\mathbf{A}^{+\bullet} \rightarrow \mathbf{B}^{+\bullet}$) or radical anion ($\mathbf{A}^{-\bullet} \rightarrow \mathbf{B}^{-\bullet}$) may be facile reactions with low activation energies.

Let us now extend the concept of "hole" or "electron" catalysis to a redox system consisting of the original reaction and an oxidant or reductant, **M**. We need not specify the nature of **M** at this stage. If we simplify the reaction system by assuming that there is no direct interaction (i.e. complexation or ion-pairing) between the reaction system, **A** → **B**, and **M**, we obtain the simple reaction profiles shown in Fig. 1.

For the case in which **M** oxidizes the substrate **A**, the reacting system crosses from the neutral (solid line) reaction profile with its high activation energy to that of the "hole-catalyzed" reaction (dashed line) as indicated by the arrows. The activation energy for the non-interacting system in the gas phase is given by:

$$\Delta E^{*}_{\text{cat.}} \simeq \Delta IP + \Delta E^{*}_{\text{hole}} \tag{1}$$

where $\Delta E^{*}_{\text{cat.}}$ is the activation energy for the catalyzed reaction, $\Delta E^{*}_{\text{hole}}$ that for the "hole-catalyzed" equivalent and ΔIP is the difference in ionization potentials between **A** and **M** (i.e. the energy needed to transfer an electron from **A** to **M** in the hypothetical non-interacting system). Clearly, both $\Delta E^{*}_{\text{hole}}$ and ΔIP should

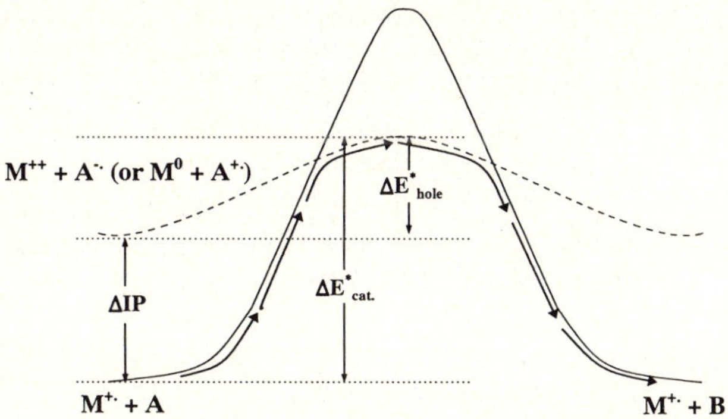

Fig. 1. Schematic energy diagram illustrating the curve-crossing model for electron-transfer catalysis. The ground state switches from the $M^{+\cdot} + A$ state to the electron-transfer ($M^{++} + A^{-\cdot}$ or $M^0 + A^{+\cdot}$) state and back again along the lowest energy path. Generally, the reaction will occur with little or no symmetry in order to avoid pseudo-Jahn-Teller intersection points

be as small as possible for effective catalysis. Similarly, reductive catalysis occurs by transferring an electron from **M** to **A**. The assumption of non-interacting components is, however, particularly unrealistic for the "naked ion" reactions usually calculated, so that a second energetic contribution, the electrostatic interaction energy within the ion pair produced by electron transfer, helps to stabilize the transition state. Thus, Eq. (1) usually represents the upper limit for the activation energy of an electron-transfer catalyzed reaction.

In practice, calculations are not usually performed on systems in which both **A** and **M** are neutral. In the gas phase (and hence for calculations in vacuo), the energy required to separate charges is prohibitive, so that model systems for catalysis calculations are generally singly charged. This also has the advantage that the singly charged species generally form stable complexes with the substrate and product, so that geometry optimization of the stationary points along the reaction path is facilitated. There are also parallels in gas-phase chemistry using "naked" (unsolvated and uncomplexed) metal ions [4], which are far more reactive than neutral atoms.

In order to understand the principles involved in electron-transfer catalysis and also in order to appreciate the historical development of the subject, we must treat "hole catalysis" and electron transfer between metal atoms and ions and organic substrates before examining catalytic reactions in more detail. This review is intended to cover the basic principles involved in these three areas and to provide a conceptual framework for electron-transfer catalysis.

2 "Hole-Catalyzed" Reactions

It is now generally recognised that radical cations are usually very much more reactive than their neutral counterparts. This reactivity, however, is often based on an apparent paradox. Longuet-Higgins and Abrahamson [5] were the first to consider the application of orbital symmetry considerations to radical reactions immediately after the publication of Woodward and Hofmann's seminal communication on the stereoelectronic factors governing electrocyclic reactions [6]. They actually introduced the use of state-correlation diagrams for treating electrocyclic reactions and pointed out that, in contrast to the cation and anion systems, there is no state-symmetry-allowed ring-closure for the allyl radical. They concluded that the radical reaction is forbidden and should be more difficult than the closed-shell equivalents. Some 12 years later, Bauld and Cessac [7] discussed an extended orbital model designed to treat radical reactions and predicted symmetry rules for simple electrocyclic radical reactions. Bischof [8] pointed out the analogy between Jahn-Teller radicals and electrocyclic reactions of radicals and defined three different types of reaction with different state correlation diagrams, as shown in Fig. 2. The labelling of the types A-C corresponds to Haselbach et al.'s later scheme [9]. In Bischof's paper types A and C are swapped. The importance of these three types of reaction is that both A and B are both orbital- and state-symmetry forbidden, whereas C is both orbital- and state-symmetry allowed. There are, however, very few (or no) examples of type C reactions, so that hole-catalyzed reactions are essentially all orbital- and state- symmetry forbidden.

Haselbach et al. [9] also classified radical electrocyclic reactions in the three types shown in Fig. 2, but were the first to point out that formally state-forbidden radical ion reactions can be extremely facile because state crossings can occur at very low activation energies. The principles outlined were used to analyze the rearrangement of the quadricyclane radical cation, **1**, to the norbornadiene radical cation, **2**, a reaction that occurs at extremely low temperatures in Freon matrices [10].

Thus, the apparent paradox lies in the fact that radical and radical-ion electrocyclic reactions are all forbidden in the Woodward-Hoffmann sense because the symmetry of the singly occupied molecular orbital (SOMO) changes

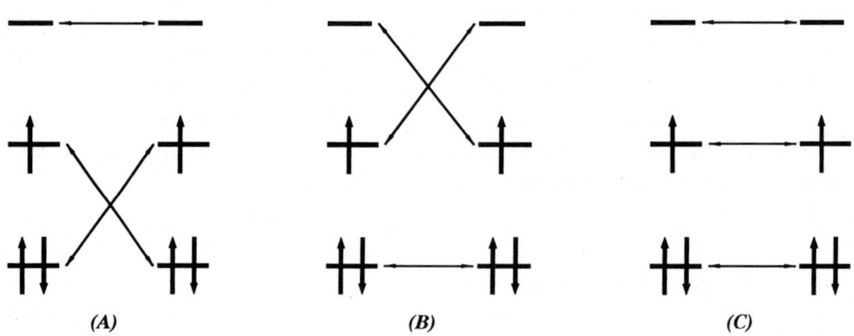

Fig. 2. The three types of radical electrocyclic reaction, as defined by Bischof [8] and Haselbach et al. [9] The notation A–C is taken from Ref [9]. Reactions of type C are essentially unknown, whereas types A and B are both orbital- and state-symmetry forbidden

from reactant to product. This is a direct consequence of the change in the number of π-bonds during the reaction. For closed-shell reactions, the allowed path always involves a change in the symmetry of the highest occupied and lowest unoccupied molecular orbitals (HOMO and LUMO, respectively), so that there is necessarily a change in the symmetry of the singly occupied MO (SOMO) in the corresponding one-electron oxidized or reduced reactions. Haselbach et al.'s analysis of this phenomenon [9] provides the basis for our understanding of "hole"-and "electron"-catalysis, although the seminal nature of his paper is often forgotten in contemporary discussions. Figure 3 shows the relevant molecular orbitals for the **1** → **2** reaction. **1** and **2** can be considered as different states of the same radical cation. They differ only in the nature of the SOMO. Quadricyclane$^{+\bullet}$, **1**, has a SOMO that is σ-bonding across the three-membered bridging bond (C_2–C_6 and C_3–C_5), but π-antibonding across the C_2–C_3 and C_5–C_6 bonds. The SOMO of norbornadiene$^{+\bullet}$, **2**, on the other hand, has the opposite phases in these bonds. The symmetrical orbital Ψ_1 doubles as the symmetrical combination of the two (C_2–C_6 and C_3–C_5) σ-bonds in **1** and the in-phase combination of the two π- bonding orbitals in **2**.

The forbidden retro-[1*s* + 2*s*]-cycloaddition can now be treated using a simple curve-crossing model analagous to the Marcus-Hush theory of electron-transfer [11]. The ground state at the quadricyclane-like geometry is the $^2B_1$ state in which Ψ_2 is singly occupied and Ψ_3 is empty. The first excited state of **1**, which lies about 50 kcal mol^{-1} above the ground state, is the $^2B_2$ radical ion that corresponds to the norbornadiene radical cation **2**. As the geometry moves towards a norbornadiene structure, the $^2B_2$ state rapidly becomes more stable as the energy of the $^2B_1$ increases.

Within exact C_{2v} symmetry, the two states may cross. The system can, however, deviate from this ideal symmetry to allow coupling between the two states so that the crossing is avoided. If we equate the $^2B_1 \rightarrow {}^2B_2$ excitation energy in **1** to the reorganization energy, λ, of Marcus theory [11], we can use the Marcus-Hush expression [Eq. (2)] to estimate the activation energy, ΔE^*, based on the calculated values for λ and the heat of reaction, ΔE (-12 kcal mol^{-1}):

$$\Delta E^* = \frac{(\lambda + \Delta E)^2}{4\lambda} \tag{2}$$

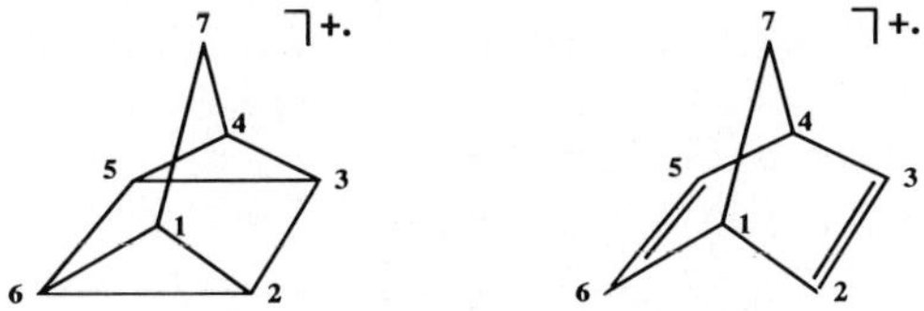

1

2

Scheme 1

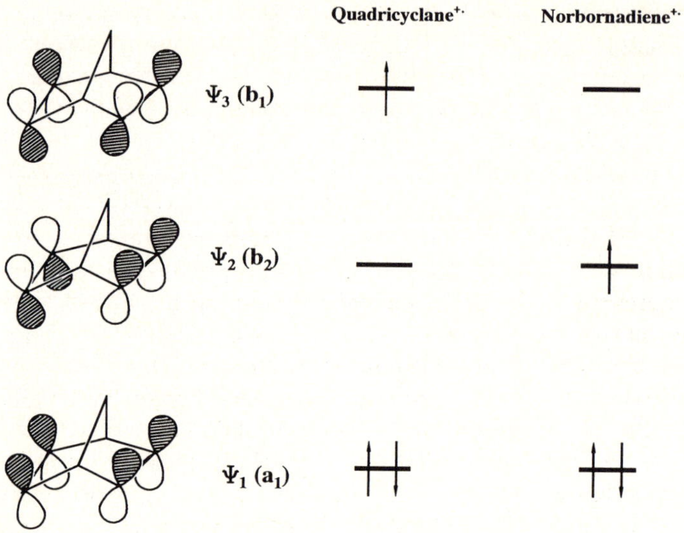

Fig. 3. π-Molecular orbitals and occupancies for the norbornadiene and quadricyclane radical cations

Note that calculated energies have been substituted for the more normal free energies of Marcus theory. Equation (2) predicts a low (7.2 kcal mol^{-1}) activation energy for the **1** $\rightarrow$ **2**. The experimental work in Freon glasses did not reveal the existence of **1**, although recent work has shown that it does have a finite lifetime in solution and that the estimated rearrangement barrier is 4.8 kcal mol^{-1} [12].

This type of simple curve-crossing interpretation can be used for many radical and radical-ion reactions and provides a pleasing link to Marcus theory. It also illustrates the solution to the apparent paradox that forbidden radical-ion reactions are so facile. A major assumption of the Woodward-Hofmann rules is that the classical activation energy for the ground state reaction is well below the lowest excitation energy, so that curve-crossing cannot occur. Radical cations and anions, however, usually have very low-lying excited states that can be stabilized along the reaction path and result in formally nonadiabatic reactions with low crossing points. It is important to note that this is a purely formal view of such reactions and that, in practice, the geometries along the reaction coordinate are strongly perturbed away from their maximum symmetry, so that ground- and excited states couple strongly and activation energies are reduced even more.

Note that one significant difference between the above "rules" for radical reactions and the Woodward-Hoffmann (W.-H) rules is that, whereas the W.-H rules for closed-shell reactions are not strictly speaking symmetry rules (they work for unsymmetrically substituted derivatives as well as for the symmetrical parent systems), the state-crossing (non-adiabatic) mechanism invoked for radical reactions does depend strictly on the symmetries of the states involved. This means that the system near a crossing point can gain energy by distorting away from the symmetrical geometry (as pointed out by Bischof in his analogy with Jahn-Teller radicals [8]), so that radical reactions of this type are likely to proceed via very asynchronous paths, unlike the classical concerted W.-H processes.

This is the case for the quadricyclane$^{+\cdot}$ to norbornadiene$^{+\cdot}$ reaction. Although the C_{2v} reaction path provides an attractive interpretational tool for understanding the progress of this reaction, its highest point represents a conical intersection at which the two relevant states have the same energy at the same geometry. This point cannot be a transition state, so that lowering the symmetry in any direction leads to a stabilization. The result is an asynchronous reaction path in which one of the two cyclopropane bonds is broken first to form the biradical-like transition state **1a**. The second bond can then break to form the norbornadiene radical cation **2**.

Radical anion reactions have been treated less thoroughly than radical cations, although the same principles should apply. Bauld et al. [13] discussed the benzocyclobutene$^{-\cdot}$ to *o*-quinodimethane$^{-\cdot}$ (**3** → **4**) rearrangement because it was thought to be a rare allowed (Type C) radical electrocyclic reaction on the basis of INDO calculations. Later calculations [14] suggest that this is in fact a normal Type B reaction with an unusually high activation energy because derivatives of both **3** [13, 15] and **4** [14] are known experimentally and their interconversion has not been observed.

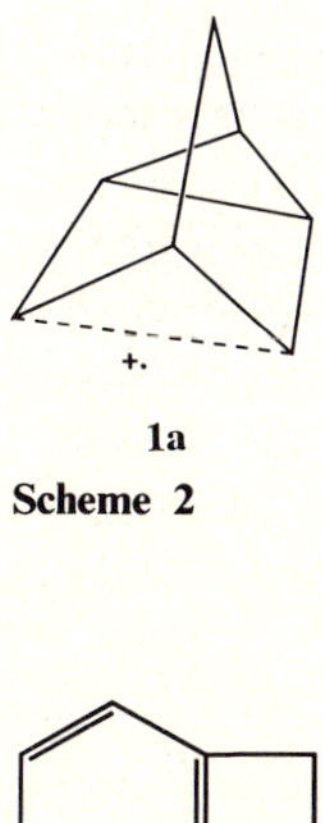

1a

Scheme 2

3 **4**

Scheme 3

3 Calculational Methods for Radicals, Radical Ions and Radical Reactions

Ab initio molecular orbital methods are well established for closed-shell reactions. Their performance has been well documented and relatively simple levels of theory give reliable results for most systems. This, however, is not the case for radicals, radical ions and radical reactions. The mostly widely used single-determinant formalism for open-shell reactions is unrestricted Hartree-Fock (UHF) [16]. Although the restricted open-shell Hartree-Fock (ROHF) formalism [17] has several advantages over UHF, it is used less frequently because it is often slower than UHF and can converge very slowly. UHF theory is not without problems. Firstly, the wavefunction is not an eigenfunction of the total spin operator, which means that higher spin states than the one being considered can mix into the wavefunction. This is usually not an extremely severe problem for UHF calculations themselves, but becomes very serious for Møller-Plesset perturbational corrections for electron correlation based on the UHF reference wavefunction (UMP*n*) [18]. Even small amounts of higher spin states (spin contamination) in the UHF wavefunction can have large effects on the UMP*n* energy [19]. Thus, MP2 calculations, which would normally be one of the methods of choice for geometry optimizations on closed-shell systems, can cause problems when based on a UHF wavefunction for open-shell systems [20]. Nevertheless, MP2 calculations are often necessary in order to avoid the so-called symmetry-breaking problem often found for ROHF and UHF calculations [21]. Perhaps the best known example of ROHF symmetry-breaking is the allyl radical, which has the symmetrical (C_{2v}) structure **5** at UHF and higher levels, but the C_s geometry **6** is found at ROHF.

The symmetry-breaking problem can, however, also be found for UHF. The complex between beryllium$^{+\bullet}$ and two ethylenes, for instance, is calculated to have two distinctly different ethylene moieties at UHF, as shown in **7**. UMP2-optimizations give the C_{2v} structure **8** [22].

These examples are typical of the effects found in radical ion studies. Gauld and Radom [23] recently showed exactly how wide the variation in calculated properties can be at a series of different levels of ab initio theory for the isomeric $CH_3F^{+\bullet}$ radical cations. Calculated C–F bond lengths varied from 1.267 to 2.034 Å for $CH_3F^{+\bullet}$, depending on the level of calculation. Such cases are, of

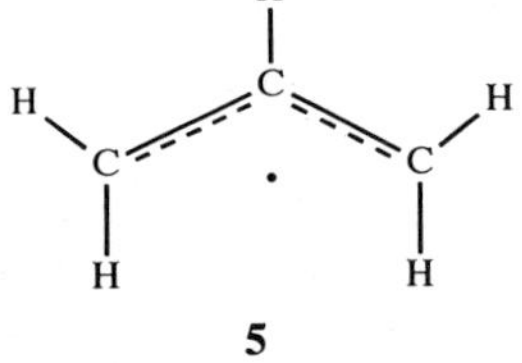

5

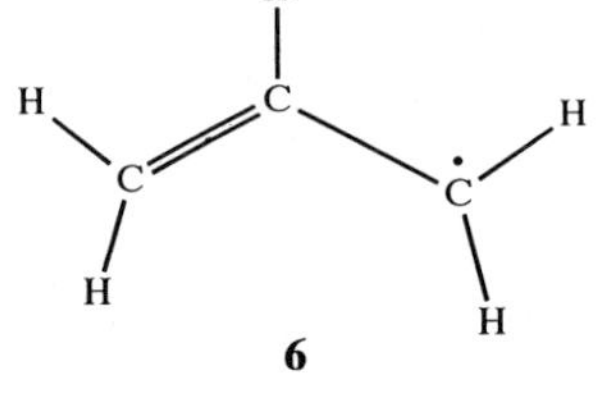

6

Scheme 4

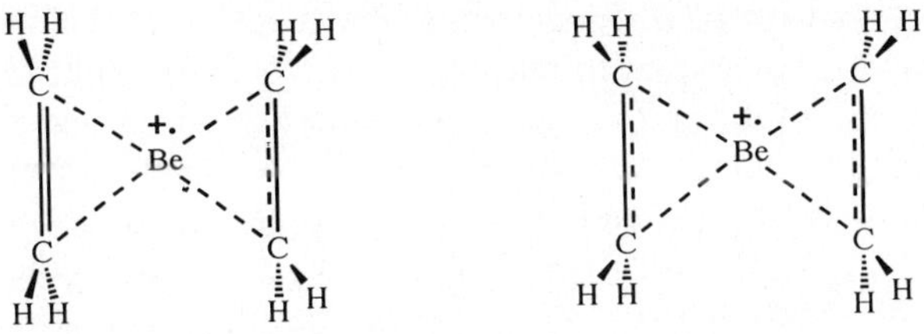

7 8

Scheme 5

course, extreme but serve to illustrate the extra difficulties often encountered in calculations for open-shell species. Neutral CH_3F, for instance, shows a variation in the calculated C–F bond length of only 0.03 Å at the same levels of theory as those used for the radical cation. Calculated energies show similar deviations. The energy difference between $CH_3F^{+\bullet}$ and its distonic [24] isomer $CH_2FH^{+\bullet}$ varies between 3.4 and 27.6 kcal mol^{-1} at the different levels of calculation. Gauld and Radom recommend the use of the so-called G2′ [25] level of theory for reliable theoretical data, although calculations of this level are often not yet practicable for extended reactivity studies. Both the $CH_3F^{+\bullet}$ study of Gauld and Radom and Wong and Radom's $C_3H_2{}^{+\bullet}$ work [26] reveal larger differences between calculated and theoretical thermochemistry than would normally be acceptable for closed-shell molecules. This discrepancy may well result from experimental errors because the highest level calculational results appear to converge to consistent values. The current situation is, however, probably best described as one in which neither experimentalists nor theoreticians can be as sure of the reliability of their results as would be the case for closed-shell systems.

A simple example serves to illustrate the difficulties. Experimentally, the ethylene radical cation is twisted by about 25° about the formal C=C bond [27]. Ab initio theory often does not reproduce this twist [28] and gives planar structures at many levels of theory for ionized olefins. MNDO semiempirical MO-theory [29] gives twisted structures for most olefin radical cations [30] but appears to overemphasise the tendency to twist, as do other NDDO-based methods. It has been suggested from an analysis of their ESR spectra [31] that many alkyl-substituted olefin radical cations are, in fact, planar. This leads to the situation where most ab initio studies give planar structures for all olefin radical cations, semiempirical methods give twisted structures and experiment suggests a mixture of planar and twisted C=C linkages. Recent density functional calculations [32, 33], however, agree well with the experimental results for a variety of olefin radical cations, suggesting that density functional theory may provide some help in the calculation of radical cation systems. Radom [34] has recently tested a large number of calculational methods, including density functional theory, for radical cations.

A critical point in the calculation of open-shell systems is the question of Jahn-Teller distortions. These are an integral component of Bischof's original qualitative treatment of radical reactions [8] and determine the structure of many cycloalkane radical ions and annulene radicals [35]. Two recent papers have treated Jahn-Teller radical cation systems with ab initio theory. Eriksson and Lunell [36] calculated the equilibrium deuterium isotope effect on the positional isomers of the Jahn-Teller distorted cyclopropane radical cation. Their results at UHF/6-31G** agree well with experimental observations, suggesting that this level of theory can deal with Jahn-Teller distortions relatively well. Krogh-Jespersen and Roth [37] have also considered methyl substituted cyclopropane radical cations. Their optimizations at the UHF level give larger distortions than those found using UMP2, a trend that has also been observed [33] for the norbornadiene/quadricyclane radical cation rearrangement system. Generally, UHF energy differences between alternative Jahn-Teller structures are also larger than those found using Møller-Plesset corrections for electron correlation. This may be a reflection of the symmetry-breaking tendencies of UHF. It is important to note here that Jahn-Teller radicals often give very significant spin-contamination of the UHF wavefunction, so that perturbational theories such as MPn based on the UHF reference wavefunction may not be reliable [19]. Eriksson et al. [32] have indicated that they plan to study Jahn-Teller radicals with density functional theory but our preliminary results [33] indicate that the lack of non-dynamic correlation in density functional theory may lead to an underestimation of the geometrical and energetic effects of Jahn-Teller distortions.

One final computational aspect of radical ion studies is the treatment of radical anions. The use of the LCAO approximation and finite basis sets means that molecular orbital calculations do not allow the dissociation of an electron even if it is unbound. This effect can be used to simulate the effect of condensed phases by trapping the electron within the confines of the basis set [38], but can also lead the unwary to false conclusions about radical anions. Consider, for instance, the acetonitrile radical anion, $CH_3CN^{-\bullet}$. Experimentally, this radical anion is a very weakly bound C_{3v} species in the gas phase. The extra electron is bound to neutral acetonitrile by interaction with its permanent dipole. This results in a very diffuse unpaired electron density akin to that in Rydberg states of neutral molecules [39]. The extra very weakly bound electron does not affect the structure of the radical anion, which is therefore that found for neutral acetonitrile. In crystalline acetonitrile, on the other hand, monomeric $CH_3CN^{-\bullet}$ has a nonlinear CCN linkage, indicating that the extra electron occupies a valence-like orbital [40]. The reason for this structural dichotomy is that very diffuse species such as Rydberg states or dipole-bound radical anions are strongly destabilized in condensed phases, where their electron density overlaps strongly with that of surrounding molecules, so that compact valence-like states become more favorable. These changes are reflected in calculations. The behavior of a radical anion calculation is very often determined by the extent of the most diffuse orbitals in the basis set because the extra electron is unbound and

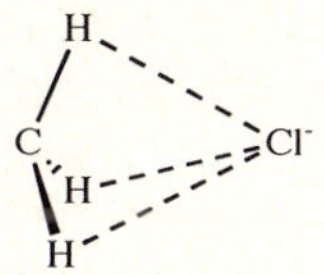

9

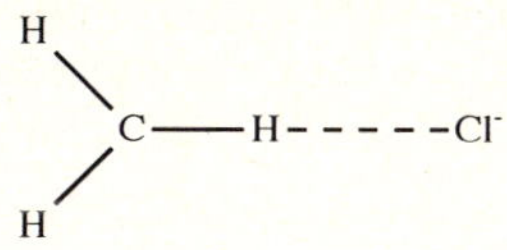

10

Scheme 6

therefore dissociates as far as the basis set allows. This effect can be used to tune a basis set to behave like a condensed phase [38] even though a calculation including Rydberg orbitals would give a completely different result. This effect is often compounded by the fact that ab initio SCF methods underestimate electron affinities consistently because they overestimate electron-electron repulsion. The calculations therefore make the extra electron even more weakly bound than it is in the real system.

One radical anion reaction system that has been examined repeatedly is the dissociation of the chloromethane radical anion into the methyl radical and a chloride ion. The ground state of the radical anion near the CH_3Cl geometry is a dipole-bound species. The first theoretical treatment of this system used an STO-3G basis set augmented with Rydberg orbitals and found a significant "derydbergidization" barrier to dissociation [41]. This barrier, which exists in the gas phase, is caused by the crossing of the dipole-bound state, whose energy follows the Morse curve for the C–Cl dissociation, and the purely dissociative (at this level of theory) curve for the σ^* valence-like radical anion. Later work using larger basis sets at the UHF and UMP2 levels confirmed this interpretation and suggested that the $CH_3Cl^{-\bullet}$ ion should dissociate without a barrier in condensed phases [38, 42]. In fact, UMP2 optimizations with the 6–31 + G* basis set [43] reveal a shallow minimum **9** on the C_{3v} dissociation path corresponding to a chloride ion weakly bound to the backside of a slightly nonplanar methyl radical [44]. The minimum energy structure for $CH_3Cl^{-\bullet}$ is, however, found to be the C_{2v} complex **10**. One further aspect of this system is that simulations of electron transfer from $Li^\bullet$ in which the lithium basis set contains only core orbitals [45] suggest that the ion pair $CH_3Cl^{-\bullet}Li^+$ has a bound minimum energy structure, in contrast to the results obtained for the isolated radical anion.

4 Radical Ion Reaction Mechanisms

4.1 Cycloadditions

The calculations of Pabon and Bauld on the hole-catalyzed ethylene dimerization were among the first ab initio studies on radical cation reactions [46].

Their conclusion that the activation energy for the dimerization is low has withstood the test of time, although their UHF calculations on the structure of the cyclobutane radical cation have since been improved by Bally et al. [47], who do not find the "long bond" structure long thought to be correct. Jungwirth and Bally's further calculations [48] reveal the following mechanism for this reaction [reaction energies and activation energies (in square brackets) are shown in kcal mol^{-1}]: The ethylene/ethylene$^{+\bullet}$ complex **11** forms without activation energy and can then undergo either a facile 1,3-hydrogen shift (see below) or ring-closure to the cyclobutane radical cation **12**. The formation of the 1-butene radical cation as primary product is in accord with experimental observations in frozen matrices [49]. The hole-catalyzed ethylene dimerization has also been treated by Lunell's group [50] and by Lee et al. [51].

Alvarez-Idaboy, Eriksson and Lunell have also studied the hole-catalyzed ethylene trimerization [52]. They find a stable complex between the 1-butene radical cation and ethylene that can rearrange with an activation energy of 9.2 kcal mol^{-1} to the 1-hexene radical cation. In this case, there is apparently no 1-ethyl-tetramethylene radical cation intermediate. Once again, the results of the calculations are in accord with experimental findings.

The prototype hole-catalyzed Diels-Alder reaction between the butadiene radical cation and ethylene has also been studied by Bauld [53]. He finds strongly exothermic formation of a 1-hexene-3,6-diyl radical cation intermediate without activation energy followed by a weakly activated (activation energy $\simeq$ 2.3 kcal mol^{-1}) closure of the second C–C bond to form the cyclohexene radical cation. The reaction shows no overall activation energy relative to the

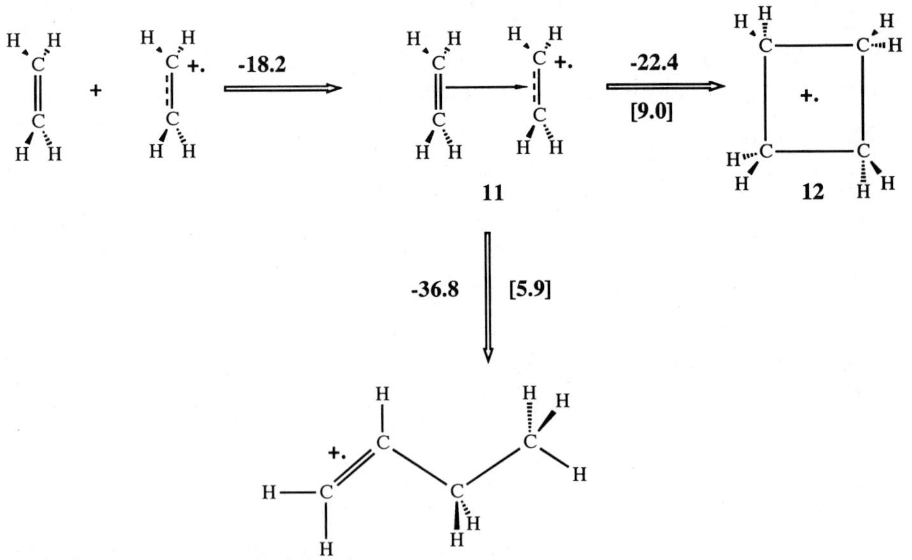

Scheme 7

starting materials. Bauld's results rationalize the observed hole-catalysis [54] of Diels-Alder reactions well, but Bally's work suggests that the details of the potential energy surface may change significantly at levels of calculation higher than the UMP2/6-31G*//(UHF/3-21G) used by Bauld.

Schwarz et al. have also studied a hole-catalyzed cycloaddition reaction involving a hetero-atom, that of the ketene radical cation with ethylene [55]. They find that the cyclobutanone radical cation is not a minimum on the potential energy surface, but rather that it undergoes α-cleavage to yield the radical cation **13**. β-Cleavage has a significant barrier and leads to a product **14** that is less stable than the cleavage products (the ketene radical cation and ethylene). The addition reaction of these two species was investigated in detail and found to involve a nucleophilic addition of ethylene to the terminal ketene carbon atom. This sort of process is typical for radical ion reactions.

4.2 Sigmatropic Rearrangements

The 1,3-hydrogen shift in propene is one of the model reactions which, despite the fact that it does not occur, is used to illustrate the principles involved in the W.-H. treatment of sigmatropic reactions. The situation for the neutral system

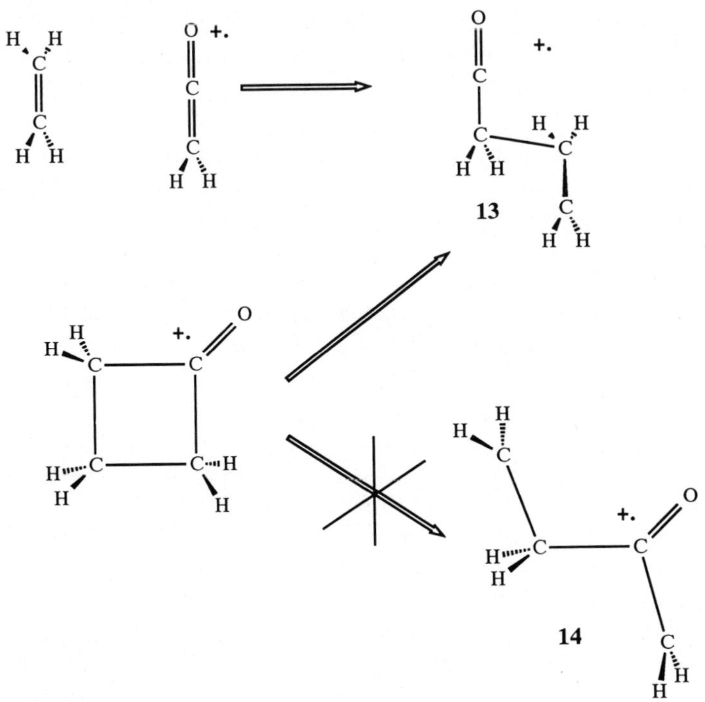

Scheme 8

was investigated by Bernardi et al. [56] using MCSCF theory. They were unable to locate a *suprafacial* transition state for the 1,3-hydrogen shift and found that the activation energy for the *antarafacial* process is essentially the dissociation energy of the C–H bond, so that the neutral sigmatropic shift hardly qualifies as a concerted reaction. This reaction was investigated for the radical cation using UHF/6-31G* and a C_s trimethylene radical cation was found to be the transition state for the suprafacial shift [57]. UMP2/6-31G* calculations on the same system revealed that this species is a shallow minimum at the higher level of theory and that the 1,3-shift is actually two consecutive 1,2-shifts [58]. Both studies give an activation energy of about 30 kcal mol^{-1} for the hydrogen shift. Nguyen et al [59] were later also able to locate an *antarafacial* 1,3-shift transition state. 1,3-Hydrogen shifts play a large role in many of the radical cation reactions, such as the dimerization and trimerization of ethylene, that have been investigated to date. Exothermic 1,3-shifts are often preferred to simple ring-closure or C–C bond-forming reactions.

Bauld [60] has also examined the hole-catalyzed vinylcyclobutane/cyclohexene rearrangement with ab initio theory. His MP2/6-31G*//(UHF/3-21G) calculations give an activation energy of 9.4 kcal mol^{-1} for this concerted reaction. Not surprisingly, the rearrangement proceeds via the same intermediate as the radical cation Diels-Alder reaction. Once again, the level of these calculations may not be sufficient for the detailed conclusions to be correct, although the general features of the calculated mechanism are probably reliable.

4.3 Electrocyclic Reactions

The ring-opening of the cyclopropane radical cation to the trimethylene radical cation and the subsequent rearrangement to the propene radical cation have been the subject of ab initio studies by Arnold et al. [61], Lunell et al. [62] and by Du, Hrovat and Borden [58]. The latter authors find a conrotatory ring-opening (activation energy 36 kcal mol^{-1}) leading to a bifurcation point for rearrangement to one of the two possible propene isomers ($C^1H_2=C_2H–C^3H_3$ and $C^1H_3–C^2H=C^3H_2$). The activation energy for the isomerization from the intermediate trimethylene radical cation to the propene radical cation is estimated to be less than 0.2 kcal mol^{-1}. All studies on this system[57, 58, 59, 61, 62] agree that the trimethylene radical cation is either not a minimum or occupies a very shallow energy well. In any case, the theoretical prediction is that the trimethylene radical cation should not be observable.

This contrasts to the situation for the oxirane radical cation, for which the nature of the species observed experimentally in frozen matrices was long disputed [63]. Tomasi and his group [64] first reported calculations on the fate of the oxirane radical cation produced by vertical excitation. Later calculations [65–67] all agree that the C–C ring-opened product **15** is formed from the oxirane radical cation with a low activation energy and that this is the structure observed in frozen matrices. Remarkably, the prototype W.-H. electrocyclic

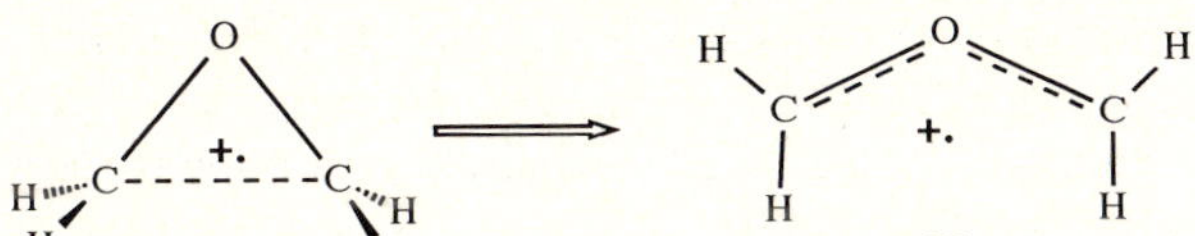

15

Scheme 9

reaction, the ring-opening of cyclobutene to 1,3-butadiene, has only been treated by semiempirical theory as a hole-catalyzed reaction [68].

5 Electron-Transfer from Metal Ions

It is tempting to relate the thermodynamics of electron-transfer between metal atoms or ions and organic substrates directly to the relevant ionization potentials and electron affinities. These quantities certainly play a role in ET-thermodynamics but the dominant factor in "inner sphere" processes in which the product of electron transfer is an ion pair is the electrostatic interaction between the product ions. Model calculations on the reduction of ethylene by alkali metal atoms, for instance [69], showed that the energy difference between the $M^0{:}C_2H_4$ ground state and the $M^+{:}C_2H_4^{-\bullet}$ electron-transfer state can be described well by a very simple equation that only considers the ionization potential of the metal, the electron affinity of the ethylene and the electrostatic ion-pairing energy:

$$\Delta E_{\mathrm{red}} \simeq IP_M^I - EA - \frac{k}{R_{MC}} \tag{3}$$

where ΔE_{red} is the electron-transfer energy, IP_M^I the first ionization potential of the metal, EA the electron affinity of the substrate ethylene, k a constant and R_{MC} the distance from the ethylene carbon atoms to the metal. the k/R_{MC} term is a simple point-charge representation of the electrostatic ion-pairing energy. This last term is the most significant component in determining the electron-transfer energy for these systems, so that the smallest alkali metal, lithium, is found to reduce ethylene best in the gas phase. This study has been extended to the reaction of ethylene with monocations of the Group II and Group XII metals [70]. These species have the intriguing possibility that they can both oxidize and reduce an organic substrate.

Contrary to expectations, the reductive process leading to the ion-pair **18** is found to be more favorable than the oxidatve path to the ion-molecule complex **17** for all Group II and most Group XII metals. Furthermore, Eq. (3) predicts the energies for the reductive process well for the Group I, II and XII metals if the constant k is adjusted for each group. These results are a little unexpected

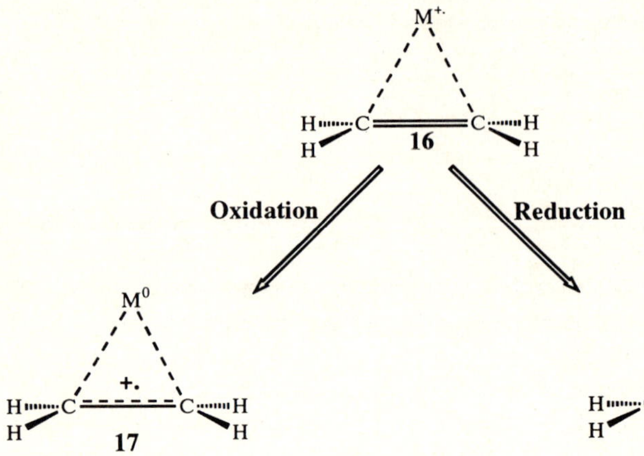

Scheme 10

because charge-separation is usually avoided in the gas phase, so that the more usual (and more stable in the absence of ion-pairing effects) reaction direction would be oxidation to **17**. The electron-transfer behavior of monocations of divalent metals is, however, of primary importance for ET-catalysis studies (see below).

6 Electron-Transfer Catalyzed Reactions

6.1 1,3-Hydrogen Shifts

The 1,3-hydrogen shift in propene was one of the first reactions to be considered in model studies on metal-ion catalyzed reactions [71]. In view of the importance of 1,3-hydrogen shifts in hole-catalyzed reactions such as the ethylene dimerization, the exact mechanism of such processes is of some importance. The monocations of Be, Mg, Ca and Zn were complexed with propene and the stationary points encountered in the course of the 1,3-hydrogen shift were optimized at UHF/3-21G with single point calculations at UMP4sdtq/6-31G*. The reaction was found to follow the two-step mechanism shown in Scheme 11. The most effective catalysis was given by beryllium, followed by zinc and magnesium. Calcium gave no catalytic effect. These results were interpreted in terms of a diagram analogous to Fig. 1, but for the oxidative process. This interpretation was based on Mulliken population analyses, which are often unreliable for split-valence basis sets, and on the rough correlation between the effectivity of the catalysis and the first ionization potentials of the metals. The later electron-transfer studies with the same metals and ethylene [70] suggest that reduction of the olefin is the more favorable process, so that the exact mechanism of the propene 1,3-hydrogen shift should be re-examined.

6.2 Cyclopropane Ring-Opening

The beryllium-cation catalyzed ring-opening of cyclopropane has been studied in some detail [72]. Two alternative products were considered; the propene:$Be^{+\bullet}$ complex **19** and ethylene + $CH_2Be^{+\bullet}$. the latter product is particularly interesting because of analogies with transition metal systems and because it provides an entry into carbene-like chemistry. Rusli and Schwarz have also considered the transfer of CH_2 from the cyclopropane radical cation to acetonitrile [73]. The two alternative products are found to have similar energies, so that there is no strong thermodynamic preference in either direction. The carbene-dissociation is, however, far more favorable kinetically than propene formation in the model naked beryllium system. The key intermediate in both reaction pathways is the berylliacyclobutane radical cation **23**. It is formed by a direct insertion of $Be^{+\bullet}$ into one of the C–C bonds of cyclopropane starting from the edge-on complex **22** (which is analogous to the cylopropane:$PdCl^+$ complex structure calculated by Siegbahn et al. [74] . This reaction is found to be activated by 29 kcal mol^{-1} (compared with Borden's value of 36 kcal mol^{-1} for the cyclopropane radical cation) [58] and to be endothermic by about 12 kcal mol^{-1}.

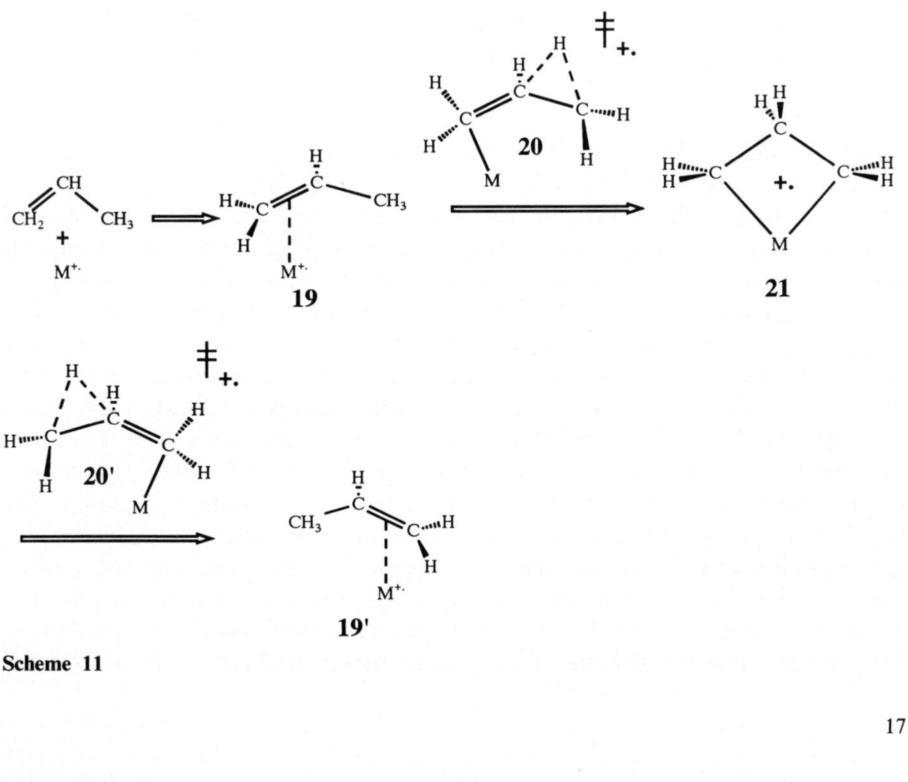

Scheme 11

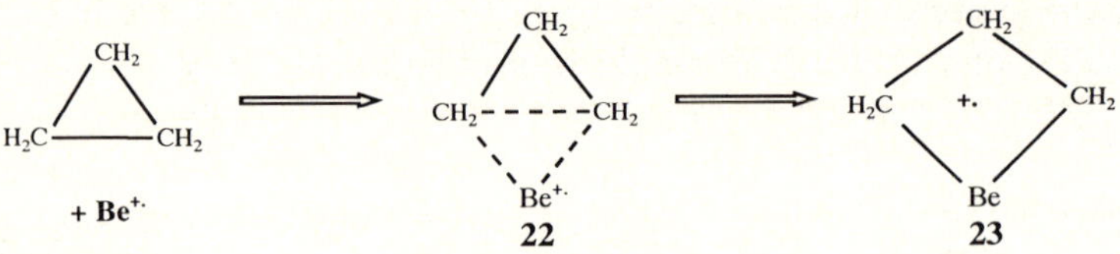

Scheme 12

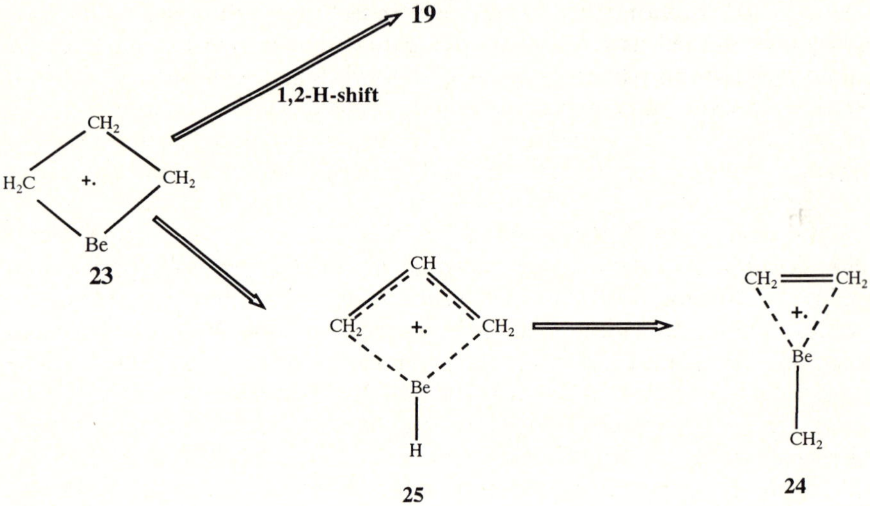

Scheme 13

The ethylene:$CH_2Be^{+\cdot}$ complex **24** is formed by a direct C–C bond-cleavage reaction that is essentially without activation energy. There are, however, two alternative pathways to the propene complex **19**. The analogous path to that found [58] for the cyclopropane radical cation is a 1,2-hydrogen shift that is found to have a barrier of about 31 kcal mol^{-1}. More favorable, however, is a hydride elimination path to give the allyl-BeH radical cation complex **25**, which then undergoes a Be–C hydrogen shift concurrently with a C–C cleavage to give **24**. The most stable species found in this study was the allyl complex **25**.

One of the surprising aspects of this and other studies using naked metal ions as models for electron-transfer catalysis are the many analogies found to known transition metal chemistry, either in the gas phase with naked ions or for complexes under more normal conditions. Clearly, such simple models as the beryllium cation cannot account for transition metal reactivity, but they do have the advantage that, because of their very simplicity, the reasons for their effects are relatively clear. The fact that $Be^{+\cdot}$ can catalyze a given reaction does not necessarily mean that, for instance, a transition metal does not use *d*-orbitals to catalyze the same reaction but it does mean that *d*-orbitals are not a prerequisite. Comprehensive studies such as that on the $Be^{+\bullet}$-catalyzed cyclopropane ring-opening are still very CPU-intensive for transition metals, so that model reactions play a useful role.

6.3 Ethylene Dimerization

The ethylene dimerization has also been investigated in detail as a beryllium cation-catalyzed reaction [22]. In this case, cyclobutane and the 1- and 2-butenes were considered as products, although others are also imaginable. The beryllium cation forms particularly stable three-electron bonded complexes with ethylene because the first ionization potential of the metal is very close to that of ethylene. This is a necessary prerequisite for strong odd-electron σ-bonds [75]. Remarkably, the MP4sdtq/6-31G*//(UHF/6-31G*) binding energy for the second ethylene ($-55\ \text{kcal mol}^{-1}$) is larger than that ($-45\ \text{kcal mol}^{-1}$) for the first. The extremely stable complex **8** undergoes an endothermic ($+25\ \text{kcal mol}^{-1}$) C–C bond forming reaction with a small ($28\ \text{kcal mol}^{-1}$) barrier (i.e. $3\ \text{kcal mol}^{-1}$ for the reverse reaction) to form the berylliacyclopentane radical cation **26**. **26** can then undergo a second C–C bond formation to form a $Be^{+\bullet}$-complexed cyclobutane, **27**, but this reaction is found to be endothermic ($+5\ \text{kcal mol}^{-1}$) with a very small ($1\ \text{kcal mol}^{-1}$) barrier for the reverse reaction. The remaining products are reached via a common intermediate, the 1-beryllia-2-methylcyclobutane radical cation **29**, which is the product of a 1,2-hydrogen shift from **26** (activation energy $27\ \text{kcal mol}^{-1}$). **29** can undergo three alternative 1,2-hydrogen shifts to form the $Be^{+\bullet}$ complexes of 1- and cis and trans 2-butenes. The lowest calculated barrier ($16\ \text{kcal mol}^{-1}$) leads to 1-butene, in accord with the fact that this is the primary product of ethylene dimerization on Pd^{+}-doped zeolites [75]. The Natural Population Analysis (NPA) procedure of Weinhold's group [76] was used in the above study to determine the nature of the wavefunction at various points along the reaction path. It revealed that this reaction is actually an electron-catalyzed process (i.e. the reductive ET pathway shown above is operative). The observed reactivity is actually that of the C_4H_8 radical anion system assisted by strong electrostatic coordination effects to the metal dication produced by electron-transfer.

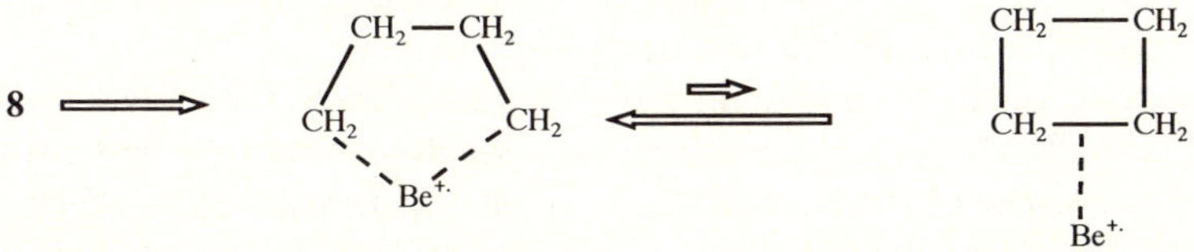

27 28

Scheme 14

27 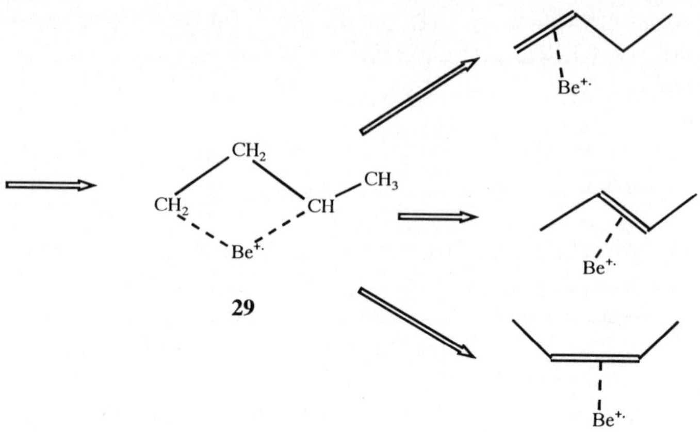

Scheme 15

6.4 C–C Bond Activation and Cycloalkane Ring-Opening by Transition Metal Atoms

Cycloalkane ring-opening reactions can be regarded as a special case of C–C bond activation. Siegbahn and Blomberg [77] have considered both types of reaction in the third in a series of papers on interactions between transition metal atoms and organic substrates. The first [78] is concerned with methane activation and the second [79] with ground state metal atom: ethylene complexes. Although they do not discuss ET-catalysis directly, their results are nonetheless important in the current context. They have considered C–C cleavages with ethane, cyclopropane and cyclobutane as substrates for the neutral atoms of the first and second row transition metals.

The results allow a rationalization of the behavior of the different metals and for the three different alkanes, but the authors do not treat the ET-aspect of the catalysis explicitly. Further examination of the transition states found in this study would be of interest.

6.5 Oxirane Ring-Opening

Model studies on the $Be^{+\bullet}$-catalyzed oxirane ring-opening have also been performed [70]. These suggest that C–O bond cleavage occurs (in contrast to the radical cation reaction) and that the $Be^{+\bullet}$: acetaldehyde complex **31**, reached via a 1,3-hydrogen shift from the ring-opened complex **30**, is the thermodynamically most stable product. Two alternative dissociations starting from **30**, however, are kinetically more favorable. C–C cleavage gives the carbene/formaldehyde complex **32**, whereas C–O bond rupture gives the $BeO^{+\bullet}$: ethylene complex **33**. The activation energies for these last two reactions are found to be

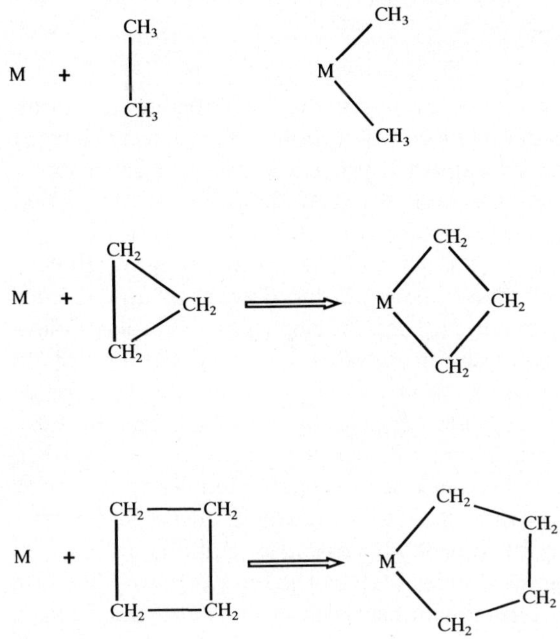

Scheme 16

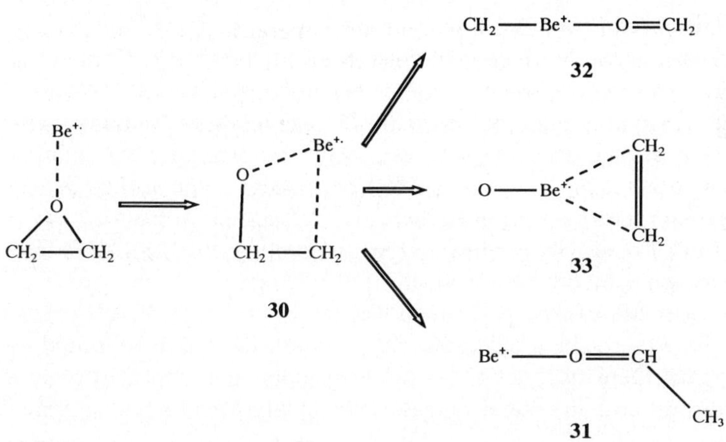

Scheme 17

low (4–5 kcal mol^{-1}) and very similar. This process is again found to be catalyzed by reductive electron-transfer. Oxirane radical anion ring-openings are known for phenyl-substituted systems with alkali metals in ether solvents [80].

7 Summary and Outlook

Although not strictly electron-transfer catalysis, the lowering of activation energies for radical or triplet reactions by complexation to a positively charged species may play a role in many ET-catalyzed processes. This effect can sometimes be very significant, as has been shown for the addition of the methyl radical to ethylene [81], 1,2-halogen shifts in β-haloalkyl radicals [82], the oxidation of methane to methanol [83, 84] and the epoxidation of ethylene [83, 85]. This electrostatic effect (it can be reproduced without any valence orbitals on the metal [84] and can therefore only be electrostatic) certainly plays an extra role in facilitating ET-catalyzed processes. Quite generally, the role of the metal is not only the ET-process shown in Fig. 1, but also to provide electrostatic acceleration and a template effect for many of the reactions described above. These simple processes must be operative in all metal-catalyzed reactions in which the metal is or becomes positively charged. There has been a tendency in the past to concentrate on orbital (covalent) effects in catalytic studies, probably because many of the theoretical interpretations are derived from extended Hückel calculations. Studies such as the ones reviewed here are very far removed from practical experimental catalyst systems but they do have the enormous advantage that they allow only very simple effects such as electron transfer or electrostatic interactions. These effects have often been neglected, but should be considered before more complicated mechanistic interpretations are introduced.

Future calculations will become closer to experimental systems. The history of ab initio calculations on both hole-catalyzed and ET-catalyzed reactions shows that the pioneering studies have often been improved and their conclusions modified again and again as better hard- and software become available. The original work has, however, almost always the provided the impetus for the later calculations. This alone is an excellent reason to push the current methods up to, and sometimes even past, their limits. Geometry optimization is a key component in all reactivity studies, so that the best applicable level is that at which the extensive geometry optimizations that belong to an investigation of alternative reaction paths can be performed. Siegbahn and Blomberg's calculations, for instance, use excellent calculational methods but are thus limited to small model systems. Density functional and eventually semiempirical theory will play an important role in extending catalytic calculations to real systems.

8 References

1. For a review, see Bauld NL (1989) Tetrahedron 45: 5307
2. See, for instance, Schmittel M, von Seggern H (1993) J Am Chem Soc 115: 2165
3. For a recent example, see Furuzumi S, Okamoto T (1993) ibid 115: 11600

4. See, for instance, Viggiano AA, Deakyne CA, Dale F, Paulson JF (1987) J Chem Phys 87: 6544
5. Longuet-Higgins HC, Abrahamson EW (1965) J Am Chem Soc 87: 2045
6. Woodward RB, Hofmann R (1965) ibid 87: 395
7. Bauld NL, Cessac J (1977) ibid 99: 23
8. Bischof P (1977) ibid 99: 8145
9. Haselbach E, Bally T, Lanyiova Z (1979) Helvetica Chimica Acta 62: 577
10. Haselbach E, Bally T, Lanyiova Z, Baertschi P (1979) ibid 62: 583
11. Marcus RA (1956) J Chem Phys. 24: 966,979; (1963) J Phys Chem 67: 853, 2889; (1964) Ann Rev Phys Chem 15: 155; (1965) J. Chem Phys. 43: 679; Marcus RA, Sutin N (1985) Biochemica Physica Acta 811: 265; Sumi H, Marcus RA (1986) J Chem Phys 84: 4894; Hush NS (1961) J Chem Soc Trans Faraday Soc 57: 557; (1967) Prog Inorg Chem 8: 391; (1968) Electrochimica Acta 13: 1005; Newton MD, Sutin N (1984) Ann Tev Phys Chem 35; 437; Sutin N (1983) Prog Inorg Chem 30: 441
12. Ishiguro K, Khudyakov IV, McGarry PF, Turro NJ, Roth HD (1994) J Am Chem Soc 116: 6933.
13. Bauld NL, Cessac J, Chang C-S, Farr FR, Holloway R (1976) ibid 98: 4561; Bauld NL, Farr FR (1969) ibid 91: 2788
14. Lorenz M, Clark T, Schleyer PvR, Neubauer K, Grampp G (1992) J Chem Soc Chem Commun 101
15. Rieke RD, Bales SE, Hudnall PM, Meares CF (1969) J Am Chem Soc 91: 2788; Bauld NL, Farr FR, Stevenson GR (1970) Tetrahedron Lett. 9: 625
16. Slater JC (1951) Phys Rev 82: 538; Pople JA, Nesbet RK (1954) J Chem Phys 22: 571; Berthier G (1954) J Chim Phys 51: 363; Amos AT, Hall GG (1961) Proc Roy Soc London Ser A 263: 483
17. McWeeny R, Dierksen G (1968) J Chem Phys 49: 4852
18. Møller C, Plesset MS (1934) Phys Rev 46: 618; Pople JA, Seeger R, Krishnan R (1977) Int J Quantum Chem Symp 11: 149; Krishnan R, Pople JA (1978) Int J Quantum Chem 14: 91; Krishnan R, Frisch MJ, Pople JA (1980) J Chem Phys 72: 4244
19. Nobes RH, Pople JA, Radom L, Handy NC, Knowles PJ (1993) J Am Chem Soc 115: 1507; Gill PMW, Pole JA, Radom L, Nobes RH (1988) J Chem Phys 89: 7307; Nobes RH, Moncrieff D, Wong MW, Radom L, Gill PMW, Pople JA (1991) Chem Phys Lett 182: 216
20. Simandiras ED, Handy NC, Amos RD (1987) Chem Phys Lett 133: 324; Simandiras ED, Handy NC, Amos RD, Lee TJ, Rice JE, Remington RB, Schaeffer HF III (1988) J Am Chem Soc 110: 1388
21. See, for instance, Paldus, J (1990) in Self-Consistent Field (Studies in Physical and Theoretical Chemistry 70) Carbó R, Klobukowski M (eds) Elsevier, Amsterdam, p 1
22. Alex A, Clark T (1992) J Am Chem Soc 114: 506
23. Gauld JW, Radom L (1994) J Phys Chem 98: 777
24. Yates BF, Bouma WJ, Radom L (1984) J Am Chem Soc 106: 5805
25. Curtiss LA, Raghavachari K, Trucks GW, Pople JA (1991) ibid 94: 7221; Curtiss LA, Jones C, Trucks GW, Raghavachari K, Pople JA (1990) J Chem Phys 93, 2537
26. Wong WW, Radom L (1993) J Am Chem Soc 115: 1507
27. Merer AJ, Schoonveld L (1969) Can J Phys 47: 1731; Köppel H, Domcke W, Cederbaum LS, Niessen Wv (1978) J Chem Phys 69: 4252
28. Lunell S, Eriksson LA, Huang M-B (1991) J Mol Struct (THEOCHEM) 230: 263; Handy NC, Nobes RH, Werner H-J (1984) Chem Phys Lett 110: 459
29. Dewar MJS, Thiel W (1977) J Am Chem Soc 99: 4899
30. Belville DJ, Bauld NL (1982) ibid 104: 294
31. Clark T, Nelsen SF (1988) ibid 110: 868
32. Eriksson LA, Lunell S, Boyd, RJ (1993) ibid 115: 6896
33. Clark T, manuscript in preparation
34. Wong MW, Radom L (1995) J Phys Chem 99: 8582; I thank Prof. Radom for a preprint of this work.
35. See, for instance, Meyer R, Graf FA, Ha T-K, Guntard HS (1979) Chem Phys Lett 66: 65; Borden WT, Davidson ER (1979) J Am Chem Soc 101: 3771
36. Eriksson LA, Lunell S (1992) ibid 114: 4532
37. Krogh-Jespersen K, Roth HD (1992) ibid 114: 8388
38. Clark T (1984) J Chem Soc Faraday Disc 78: 203
39. Stockdale JA, Davis FJ, Compton RN, Klots GE (1974) J Chem Phys 60: 4279; Compton RN, Reinhardt PW, Copper D (1978) ibid 68: 4360
40. Williams F, Sprague ED (1982) Accts Chem Res 15: 408
41. Canadell E, Karafiloglou P, Salem L (1980) J Am Chem Soc 102: 855

42. Clark T (1984) J Chem Soc Chem Commun 93
43. Hehre WJ, Ditchfield R, Pople JA (1972) J Chem Phys 56: 2257; Kariharan PC, Pople JA (1973) Thoer Chim Acta 28: 213; Clark T, Chandrasekhar J, Spitznagel GW, Schleyer PvR (1983) J Comput Chem 4: 294
44. Tada T, Yoshimura R (1992) J Am Chem Soc 114: 1593
45. Clark T (1985) J Chem Soc Chem Commun 529
46. Pabon RA, Bauld NL (1984) J Am Chem Soc 106: 1145
47. Jungwirth P, Čársky P, Bally T (1993) ibid 115: 5776
48. Jungwirth P, Bally T (1993) ibid 115: 5783
49. Fujisawa J, Sato S, Shimokoshi K (1986) Chem Phys Lett 124: 391
50. Alvarez-Idaboy JR, Eriksson LA, Fangstrom T, Lunell S (1993) J Phys Chem 97: 12737
51. Lee T-S, Lien M-H, Sen S-F, Wu H-F, Gau Y-F, Chang T-Y (1988) J Mol Struct (THEOCHEM) 121: 47
52. Alvarez-Idaboy JR, Eriksson LA, Lunell S (1993) J Phys Chem 97: 12742
53. Bauld NL (1992) J Am Chem Soc 114: 5800
54. Belville DJ, Bauld NL (1982) ibid 104: 2665; Bauld NL, Belville DJ, Pabon RA, Chelsky R, Green G (1983) ibid 105: 2378
55. Heinrich N, Koch W, Morrow JC, Schwarz H (1988) ibid 110: 6332
56. Bernardi F, Robb MA, Schlegel HB, Tonachini G (1984) ibid 106: 1198
57. Clark T (1987) ibid 109: 6838
58. Du P, Hrovat DA, Borden WT (1988) ibid 110: 3406
59. Nguyen MT, Landuyt L, Vanquickenborne LG (1991) Chem Phys Lett 182: 223
60. Bauld NL (1990) J Comput Chem 11: 896
61. Wayner DDM, Boyd RJ, Arnold DR (1985) Can J Chem 63: 3283
62. Lunell S, Yin L, Huang m-B (1989) Chem Phys. 139: 283
63. Snow LD, Williams (1988) Chem Phys Lett 143: 521; Symons MCR, Wyatt JL (1988) ibid 146: 473
64. Cimiraglia R, Miertus S, Tomasi J (1980) J Mol Struct 62: 249
65. Bouma WJ, MacLeod J, Radom L (1971) J Am Chem Soc 101: 5540
66. Clark T (1984) J Chem Soc Chem Commun 666
67. Bouma WJ, Poppinger D, Saebo S, MacLeod J, Radom L (1984) Chem Phys Lett 104: 198
68. Belville DJ, Chelsky R, Bauld NL (1982) J Comput Chem 3: 548
69. Hänsele E, Clark T (1991) Zeitschrift für Physikalische Chemie 171: 21
70. Alex A, Clark T, manuscript in preparation; Alex A, Ph.D. Thesis Universität Erlangen-Nürnberg 1993
71. Clark T (1989) J Am Chem Soc 111: 761
72. Alex A, Clark T (1992) ibid 114: 10897
73. Rusli RD, Schwarz H (1990) Chem Ber 123: 535
74. Bäckvall J-E, Björkmann EE, Petterson L, Siegbahn PEM, Strich A (1985) J Am Chem Soc 107: 7408
75. Clark T (1988) ibid 110: 1672
76. Foster JP, Weinhold F (1980) J Am Chem Soc 102: 7211; Reed AE, Weinhold F (1983) JChem Phys. 78: 4066; Reed AE, Weinstock RB, Weinhold F (1985) ibid 83: 735; Reed AE, Weindhold F (1985) ibid 83: 1736; Carpenter JE, Weinhold F (1988) J Mol Struct (THEOCHEM) 169: 41; Reed AE, Curtiss LA, Weinhold F (1988) Chem Rev 88: 899
77. Siegbahn PEM, Blomberg MRA (1992) J Am Chem Soc 114: 10548
78. Svennsson MJ, Blomberg MRA, Siegbahn PEM (1991) ibid 113: 7076; (1992) ibid 114: 6095
79. Blomberg MRA, Siegbahn PEM, Svennsson MJ (1992) J Phys Chem 96: 5783
80. Boche G, Wintermayr H (1981) Angew Chemie 93: 923
81. Clark T (1986) J Chem Soc Chem Commun 1774
82. Onciul ARv, Clark T (1989) ibid 1082
83. Hofmann H, Clark T (1991) J Phys Chem 98: 13797
84. Hofmann H, Clark T (1994) J Am Chem Soc 111: 761
85. Hofmann H, Clark T (1990) Angew Chemie 102: 697

Received: March 1995

Photoreactions of Metal Complexes with DNA, Especially Those Involving a Primary Photo-Electron Transfer

Andrée Kirsch-De Mesmaeker[1], **Jean-Paul Lecomte**[1], **and John M. Kelly**[2]

[1] Service de Chimie Organique Physique CP 160/08, Université Libre de Bruxelles, 50 Av.F.D.Roosevelt, B-1050 Bruxelles, Belgium
[2] Chemistry Department, Trinity College, Dublin 2, Ireland

Table of Contents

J.P.L. is Research Assistant and A.K-D.M. Director of Research of the National Fund for Scientific Research (Belgium)

The photochemical reactions induced by metal complexes with DNA are reviewed, with most emphasis on polypyridyl complexes [especially those of Ru(II), Rh(III) and Co(III)], porphyrins and uranyl ions. In each case, where available, data on the binding of the complexes to DNA, the effect of such binding on the excited state lifetime and other properties, and the primary photochemical reactions are described. Particular emphasis is laid on reactions involving electron transfer. Evidence for direct oxidation of the DNA bases, production of DNA-damaging reactive species, direct hydrogen abstraction from the DNA ribose and formation of adducts between the metal complex and DNA are summarised. The use of metal complexes as photophysical and photochemical probes for studying the structure and conformation of nucleic acids and possible photo-therapeutic applications are discussed.

List of Abbreviations

bpy = 2,2′-bipyridine
bipym = 2,2′-bipyrimidine
TAP = 1,4,5,8-tetraazaphenanthrene
$diCH_3TAP$ = 2,7-dimethyl-1,4,5,8-tetraazaphenanthrene
DPPZ = dipyrido[3,2-a:2′,3′-c]phenazine
PPZ = 4′,7′-phenanthrolino-5′,6′:2,3-pyrazine
phen = 1,10-phenanthroline
HAT = 1,4,5,8,9,12-hexaazatriphenylene
TMP = 3,4,7,8-tetramethyl-1,10 phenanthroline
phi = 9,10-phenanthrenequinone diimine
DIP = 4,7-diphenyl-1,10-phenanthroline
BIQ = 2,2′-biquinoline
phen-T = 4-TEMPO-1,10-phenanthroline, (TEMPO = 2,2,6,6-tetramethyl-piperidine-N-oxyl)
dpp = 2,3-di-2-pyridylpyrazine
tpy = 2,2′:6′,2″-terpyridine
DMB = 4,4′-dimethyl-2,2′-bipyridine
tmen-AO = {10-[6-(2-*N*,*N*-dimethylaminoethyl)-methylamino]hexyl}-3,6-bis(dimethylamino)-acridine
tmen = *N*,*N*,*N*′,*N*′-tetramethylethylenediamine
en = ethylendiamine
py = pyridine
tpt = 2,4,6-tripyridyl-triazine
POQ = 5-[4-[(7-chloro-quinolin-4-yl)amino]-2-thia-butylcarboxamido]phen-anthroline
CT-DNA = calf thymus DNA
poly[d(A-T)] = double stranded polydeoxyadenylic-deoxythymidylic acid
poly[d(G-C)] = double stranded polydeoxyguanylic-deoxycytidylic acid
$MTMPyP^{4+}$ = metallo-*meso*-tetrakis(4-*N*-methylpyridiniumyl)porphyrin

1 Why Study the Binding and Photoreactions of Metal Complexes With DNA?

During the last ten years, studies of luminescence and photochemistry of polypyridyl Ru(II), Rh(III) and Co(III) complexes, porphyrins and uranyl salts, in the presence of biological macromolecules such as DNA, have been the focus of increasing research work. The interest in such coordination compounds stems from their easily tunable properties. Not only their size and shape but also their spectroscopic characteristics in absorption and emission and their photophysics and photochemistry can be easily modulated [1].

The accumulation of data on these metallic complexes in the area of solar energy conversion and storage [2], has allowed a rationalisation of the photochemistry of these compounds [3, 4]. Therefore they can now be easily applied in the context of DNA biochemistry.

In this developing field, the aims may be manifold. Metal complexes may be used for probing different structures (conformations, topologies) adopted by the nucleic acids, whose polymorphism is a well known phenomenon [5]. They can also be used as luminescent markers or labelling agents of DNA in place of using radioactive markers [6] for example when they are tethered to oligonucleotides [7, 8, 9]. Moreover complexes which can photo-cleave DNA site-specifically could find applications for mapping or footprinting experiments. Those metal compounds which produce DNA-adducts under irradiation might also mark irreversibly a targeted DNA site. The metal complexes can therefore be regarded as good candidates for preparing artificial endonucleases [10], or be used as non specific (photo)-cleaving agents for probing DNA-protein contacts.

In a quite different context, the metal complexes with DNA will certainly find interesting applications in supramolecular (photo)chemistry [11]. In this regard, DNA may be used as an interesting molecular scaffolding for the study of DNA-mediated electron transfer processes and charge migrations through the DNA double helix.

Finally these compounds may also present in the future some interest in the area of novel anti-tumor drugs, or in photochemotherapy [12]. Up to now the mechanism of action of the Pt complexes as antineoplastic agents has been extensively examined [13]. In spite of their successful applications, there is a need in compounds that are less toxic than the Pt derivatives. Therefore novel photo-active metallic compounds could offer future possibilities for cancer treatment. The fact that a drug would become active exclusively under illumination might represent an advantage.

Due to the numerous applications that have stimulated studies of interactions and reactions of metal complexes with DNA, we cannot cover all the aspects in this review. We will focus our attention on photoreactions of metal complexes with DNA. Although, obviously for carrying out photochemical reactions, it is essential to discuss the binding of these compounds to DNA. Dark reactions, on the other hand, will not be described.

2 Photo-Active Metal Complexes

The photochemistry of metal complexes has been extensively studied since the 1960s and has been the subject of a number of monographs and reviews [1, 11, 14–19]. Systematic investigations have concentrated on d^3 or d^6

transition metal complexes because of the nonlability of the complexes and their favourable photophysical properties. However such stability is not a prerequisite, so long as the excited state reaction is able to take place within the lifetime of the complex.

2.1 Excited States

The initial excitation will in general lead to a vibrationally and electronically excited state of the same spin multiplicity as the ground state. Subsequent relaxation may lead by internal conversion to lower lying states of the same multiplicity (including return to the ground state) or through intersystem crossing to states of different spin multiplicity. The nature of the excited states of many metal complexes has been well-characterised by both steady state and transient absorption, emission and resonance Raman spectroscopic methods. The excited states can be conveniently classified as metal-centred (MC), ligand centred (LC) and charge transfer (either from metal to ligand (MLCT) or ligand to metal (LMCT) (Fig. 1). Of course, such a nomenclature is an approximation as metal-centred molecular orbitals will have some ligand character (and vice versa) depending on the extent of covalent character of the particular bond involved. The excited states so formed may undergo a number of chemical processes, which correlate well with the class of the excited state involved. These are discussed in further detail in Sects. 2.2 to 2.4.

As well as returning to the ground state by radiative or radiationless processes, excited states can be deactivated by electronic energy transfer. The principal mechanisms for this involve dipole-dipole interactions (Förster mechanism) or exchange interactions (Dexter mechanism). The former can take place over large distances (5 nm in favourable cases) and is expected for cases where there is good overlap between the absorption spectrum of the acceptor and the emission spectrum of the donor and where there is no change in the spin

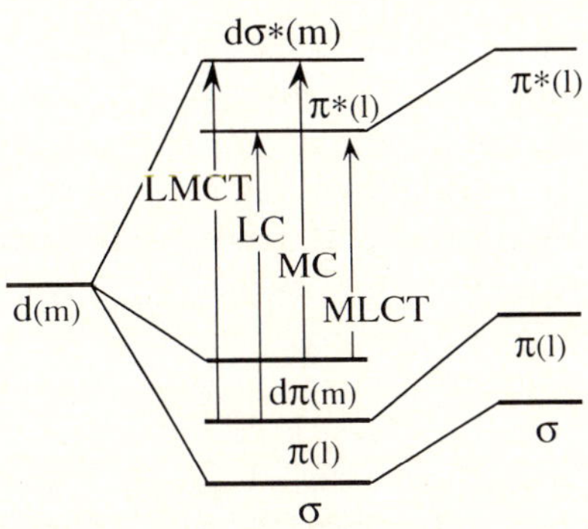

Fig. 1. Schematic diagram of molecular orbitals and electronic transitions in the case of most polypyridyl Ru(II) complexes

multiplicity. It has been well established for condensed ring aromatic and heteroaromatic compounds, such as porphyrins. For the exchange mechanism to be effective it is necessary that there is orbital overlap. This is the mechanism found for most metal complex excited states studied so far. Changes of spin state is permitted and this process therefore provides a useful route to excited states which cannot be reached directly by light absorption from the ground state (such as triple states of species with singlet ground states). Another important example is the deactivation of triplet states by ground state triplet oxygen with the resultant formation of their ground states and the formation of oxygen in a singlet excited state. This latter species is highly reactive and it is well-known to cause oxidative damage to DNA.

$$^{3}D + {}^{3}O_2 \longrightarrow {}^{1}D_0 + {}^{1}O_2 \qquad (1)$$

2.2 Ligand Dissociation and Photosubstitution

This is expected to be favoured for metal-centred excited states: for example, in *d-d** states of d^3 or d^6 complexes, where excitation often involves promotion of an electron from an essentially non-bonding orbital to one with appreciable sigma antibonding M-L character (e.g. in $Cr(NH_3)_5Cl^{2+}$, Eq. 3). The net effect is lengthening of the M-L bond, which predisposes the complex to dissociation or associative substitution. The incoming ligand is often the solvent (e.g. as in Eq. 3) or counterion of an ion pair (Eq. 4).

$$Cr(NH_3)_5Cl^{2+} + H_2O \xrightarrow{\Delta} Cr(NH_3)_5(H_2O)^{3+} + Cl^- \qquad (2)$$

$$Cr(NH_3)_5Cl^{2+} + H_2O \xrightarrow{h\nu} Cr(NH_3)_4(H_2O)Cl^{2+} + NH_3 \qquad (3)$$

$$Ru(bpy)_3^{2+} + 2X^- \xrightarrow{h\nu} Ru(bpy)_2X_2 + bpy \qquad (4)$$

(bpy = 2,2′-bipyridine, see Fig. 2)

The photochemical product is often different from that formed by heating the reaction mixture. For example thermal aquation of $Cr(NH_3)_5Cl^{2+}$ yields $Cr(NH_3)_5(H_2O)^{3+}$ (Eq. 2), whereas photochemically ammonia aquation (Eq. 3) predominates over chloride substitution. Such observations for Cr(III) complexes are consistent with rules proposed by Adamson [20]. These predict the labilised ligand – generally the stronger field ligand on the axis with ligands of average lowest field – and have been explained by various theoretical approaches. It should be emphasised that it is not necessarily the excited state initially formed which is the reactive species. Such reactions leading to photosubstitution can take place from lower MC states (possibly of differing spin multiplicity) than those initially excited, or MC state populated from CT (charge transfer) or LC states. Bands corresponding to transitions to CT states generally have significantly higher absorption coefficients than those involving MC states

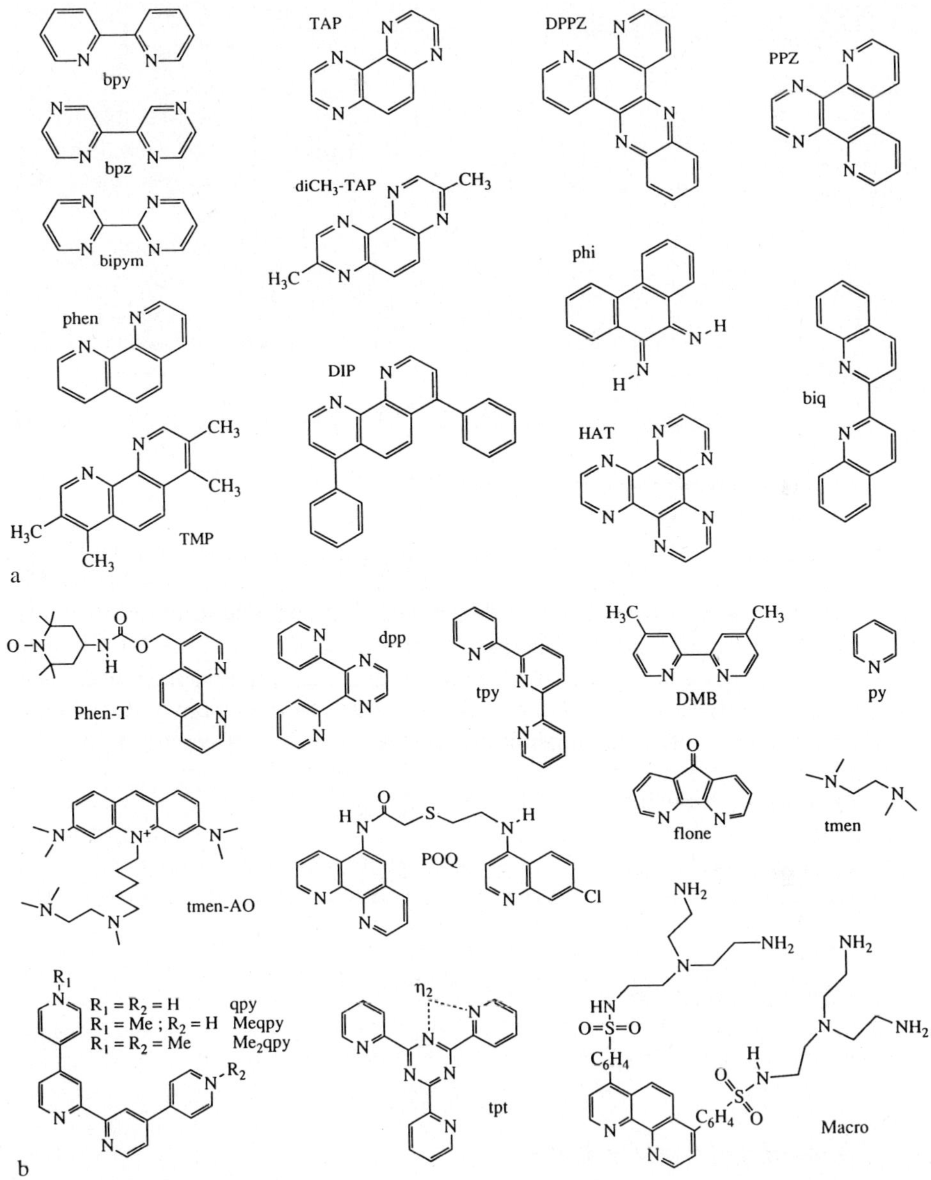

Fig. 2. Ligands corresponding to the different complexes of Table 2 and cited throughout the review. See table of abbreviations

and as a result the latter are often masked. Reactive MC states can also be formed from other states by thermal activation [e.g. MLCT states in Ru(II)polypyridyls (see below) or lower-lying less reactive MC states in Cr(III) complexes].

2.3 Homolytic Bond Cleavage

Such reactions are commonly found as a result of the decomposition of charge transfer excited states. For example, while excitation of the MC bands of $Co(NH_3)_5X^{2+}$($X = Cl, Br, I$) leads to photosolvation and the formation of $Co(NH_3)_5(OH_2)^{3+}$ and $Co(NH_3)_4(OH_2)X^{2+}$, shorter wavelengths yield the LMCT state which decomposes into Co(II) ions and halogen atoms. The quantum yield for the reaction is found to depend on the excitation energy, indicating a role for the initially formed radical pair (Eq. 5). This may reform the starting complex (Eq. 6) or decompose to the redox products stabilised by the solvent or some other species (Eq. 7). The Co(II) complexes eventually decomposes (Eq. 8).

$$Co(NH_3)_5X^{2+} \xrightarrow{h\nu} [Co(NH_3)_5^{2+}, X] \tag{5}$$

$$[Co(NH_3)_5^{2+}, X] \longrightarrow Co(NH_3)_5X^{2+} \tag{6}$$

$$[Co(NH_3)_5^{2+}, X] + S \longrightarrow [Co(NH_3)_5S^{2+}, X] \tag{7}$$

$$[Co(NH_3)_5S^{2+}, X] \longrightarrow Co^{2+} + 5NH_3 + S + X \tag{8}$$

Similarly many organometallic compounds undergo homolytic cleavage upon photoexcitation [16]. For example compounds such as $Co(CN)_5CH_2Ph$, alkyl derivatives of Co(III)macrocylic compounds, including coenzyme B12, or metallocenes Cp_2MR_2 ($M = Ti, Zr, Hf, Cp = $ Cyclopentadienyl) yield alkyl radicals, which are often responsible for further reactions involving hydrogen abstraction, addition to organic moieties, or trapping of oxygen.

Bands associated with the formation of ion pairs are observed. Excitation within these Charge Transfer to Ion (CTTI) bands causes reduction of metal centre (e.g. Eq. 9). Similarly short wavelength U.V. excitation of pentamino-Co(III) complexes in organic solvents is believed to cause solvent oxidation (Eq. 10).

$$[Co^{III}(NH_3)_6^{3+}, I^-] \xrightarrow{h\nu} [Co^{II}(NH_3)_6^{2+}, I\cdot] \tag{9}$$

$$[Co^{III}(NH_3)_5X^{2+}]S \xrightarrow{h\nu} [Co^{II}(NH_3)_5X^+, S\cdot^+] \tag{10}$$

2.4 Photo-Induced Electron Transfer

Excited states are both better oxidising and better reducing agents than their ground states. To a first approximation the oxidation and reduction potentials can be calculated as follows:

$$E^0(M^+/M^*) = E^0(M^+/M) - E^* \tag{11}$$

$$E^0(M^*/M^-) = E^0(M/M^-) + E^* \tag{12}$$

($E^* = $ 0–0 transition energy of the reactive excited state in eV).

For example the redox potentials (E^0 versus NHE) of the intensively studied $Ru(bpy)_3^{2+}$ in the systems ($Ru(bpy)_3^{3+}/Ru(bpy)_3^{2+}$) and ($Ru(bpy)_3^{3+}/Ru(bpy)_3^{2+*}$) are 1.26 and −0.87 V respectively while $E^0(Ru(bpy)_3^{2+}/Ru(bpy)_3^{+})$ and $E^0(Ru(bpy)_3^{2+*}/Ru(bpy)_3^{+})$ are − 1.35 and 0.78 respectively. Variation of the ligand (Fig. 2) can have a marked effect on these redox potentials providing a very convenient tunability of their oxidising or reducing ability. Studies of the photo-oxidation or photo-reduction of a wide range of organic and inorganic substrates have been reported [1].

With some metal complexes, e.g. $Fe(CN)_6^{4-}$, where a clear CTTS (charge transfer to solvent) band is evident, photoexcitation can cause direct photoionisation and the creation of the solvated electron.

$$Fe(CN)_6^{4-} \xrightarrow{h\nu} Fe(CN)_6^{3-} + e^-(aq) \tag{13}$$

2.5 Photo-Induced Hydrogen-Atom Transfer

It is well-known that many organic excited states (e.g. the triplet state of benzophenone) can effectively abstract hydrogen atoms from organic compounds such as alkanes and alcohols. This behaviour is not commonly found for metal-containing compounds – a notable exception being the lowest excited state of uranyl ion which abstracts H atoms from alcohols, sugars etc., with the resultant formation of free radicals and U(V) compounds. Recent work has shown that it is very effective in inducing strand breaks in DNA (see Sect. 8).

3 Binding to DNA

3.1 DNA Structure and Conformation [21–25]

DNA is a polyelectrolyte composed of deoxy-ribonucleotide monomers (mononucleotides), covalently attached to each other (Fig. 3). They consist of a β-D-2′-deoxyribose sugar bound to a negatively charged phosphate via the C'_5 and C'_3 hydroxyl and to one of the four possible bases via the C'_1 hydroxyl (glycosidic bonds); the four bases are: adenine (A), guanine (G) (both purine bases), cytosine (C) and thymine (T) (both pyrimidine bases). The purines are linked to the sugar via the N_9 and the pyrimidines via the N_1. The C'_3 hydroxyl of one nucleotidic unit is linked to the C'_5 phosphate of the adjacent nucleotide (phosphodiester bond). The direction of the polynucleotide is defined as running from the 5′ to the 3′ sugar carbons along the phosphodiester bond.

Watson and Crick showed in 1953, using X-ray diffraction data of hydrated DNA fibres [26], that B-DNA, the most commonly encountered form, corresponds to a right-handed double-stranded helix (Fig. 4). Two polynucleotide

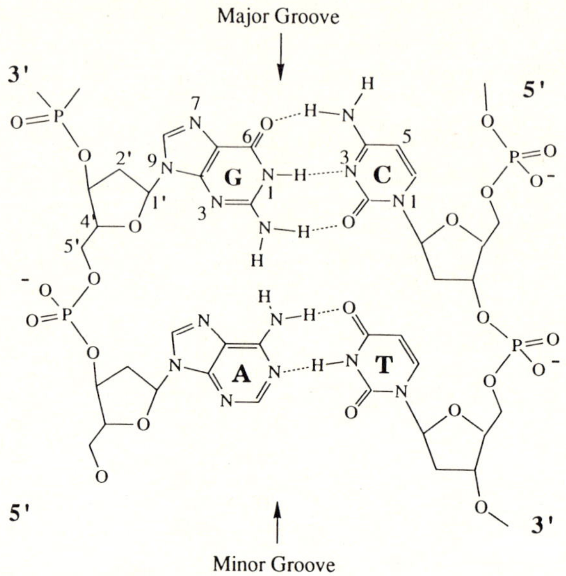

Fig. 3. Primary structure of DNA [Adapted from Dickerson RE (1983) Scientific American 6: 86]

strands are bound together by H-bonds formed between the complementary bases of each strand. The double strands are coiled around a common axis to form a right handed double helix. The two specific, complementary, hydrogen-bonded, base-pairs are: adenine-thymine and guanine-cytosine. These base-pairs are roughly planar, and are stacked one above the other inside the helix, perpendicular to the helical axis. The resulting double helix can be characterised by a coat/core structure of hydrophilic backbone of ribose phosphate around the hydrophobic pile of stacked bases. Successive base-pairs are separated by 3.4 Å and rotate an average of 36 ° around the helical axis, so that the structure repeats itself after ten residues on each chain. The sugars linked via the C'_1 carbon to the two complementary bases are not disposed symmetrically around the helical axis. This asymmetry induces the formation of two grooves of different sizes (Fig. 4), with almost the same depth but different widths (of 11.7 and 5.7 °A); they are called major and minor grooves respectively.

From single-crystal X-ray analyses, it is clear that double-stranded polynucleotides can adopt a wide variety of conformations (known as DNA polymorphism) which, however, can be classified into three main categories (Fig. 4): the right handed B-DNA as described above; another right-handed helix, called A-DNA; and a left-handed helix, called Z-DNA, because of the zig-zag structure of the sugar phosphate backbone. These three DNA differ by the conformation of the furanose ring (C'_2 *endo* or C'_3 *endo*) and the orientation of the base around the glycosidic bond (*anti* or *syn*). For example in the B-form described above, the glycosidic orientation is *anti* and the sugar conformation is predominantly

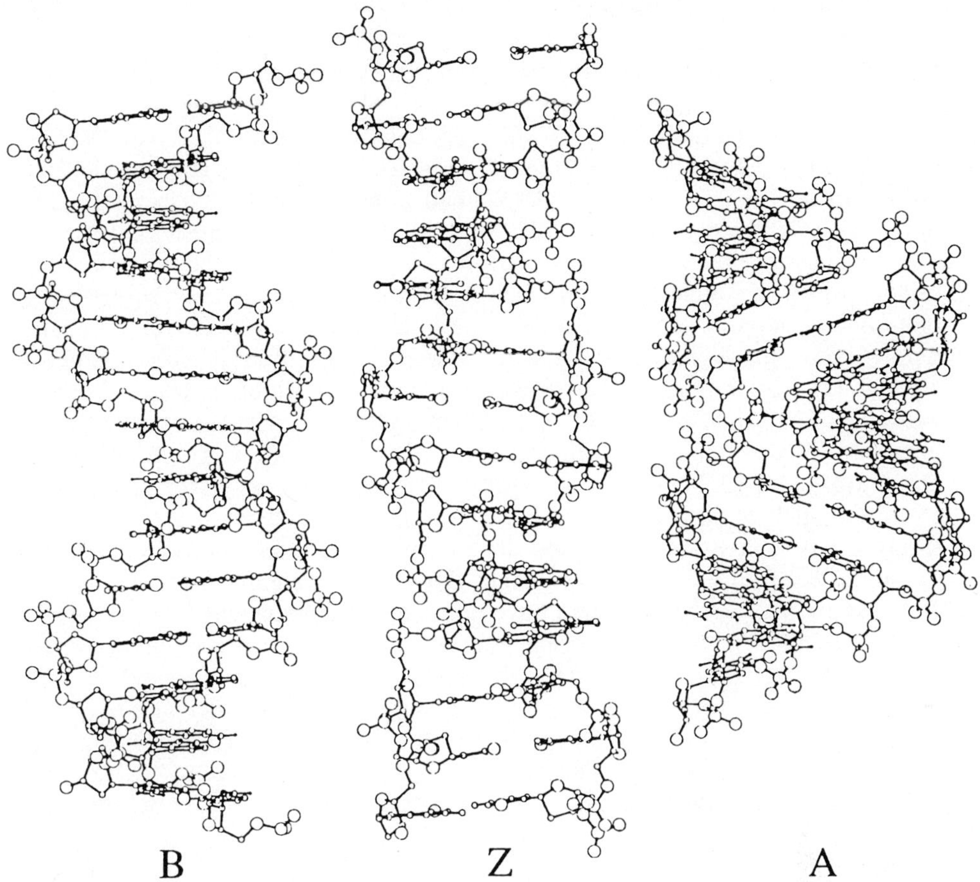

Fig. 4. Representation of B-, Z-, and A-DNA. Graphics display was performed with QUANTA (MSI) on Silicon Graphics 4D25G Personal Iris workstation, hard copies obtained with Tektronix RGBII. LEDSS, Université J. Fourier, Grenoble, France

a C'_2 *endo* conformation. Moreover the three DNA in Fig. 4, differ also by the external topology. Instead of narrow minor grooves and opened major grooves typical of B-DNA above, the A-form helix found in DNA/RNA duplex or in RNA, is distinguishable from B-DNA by a wide and shallow minor groove and a narrow and deep major groove.

Z-DNA, discovered in 1979, with the determination of the structure of the double-stranded hexamer "*CGCGCG*" has a left-handed conformation but is not just a mirror image of B-DNA [27, 28] (Fig. 4). Indeed the major groove is wide and shallow while the minor groove is narrow and deep. This Z-DNA structure is mostly adopted by alternating cytosine/guanine sequences at high ionic strength or in the presence of polyvalent ions such as $Co(NH_3)_5^{3+}$.

3.2 DNA-Ligand Interactions

DNA can interact with proteins, drugs, carcinogens, mutagens and dyes, according to different modes of binding [21].

The binding of a molecule to DNA is very complex, therefore it is worthwhile to recall the different types of "forces" or processes that are responsible for a decrease of the free energy of DNA-small molecules systems: (1) hydrogen bondings between interacting species that have hydrogen bond accepting or donating groups and the bases heteroatoms can take place via the major or the minor groove; (2) different types of electrostatic interactions may be present: monopole/monopole (between cations and the negative phosphate groups), dipole/dipole, dipole/induced dipole and also higher multipole/multipole interactions; (3) London dispersion forces (instantaneous dipole/induced dipole) between a molecule and DNA; (4) entropy increases which is particularly important in the hydrophobic interaction between the base stacking and aromatic molecules [29], or more generally speaking in the displacement of the hydration shell of both partners when a molecule interacts with DNA. Structural aspects such as the shapes and dimensions of the grooves (which can lead to steric hindrance, destabilizing the DNA-ligand complex) or repulsive electrostatic interactions can also have non-negligible influence.

All these factors induce external molecules to interact in various fashions with DNA. The different "interaction modes", commonly considered in the literature, are: surface binding in the major or minor groove, intercalation, and external binding in the atmosphere of ions of the DNA polyelectrolyte (Fig. 5).

If the ligand has an aromatic portion (normally corresponding at least to three fused six-membered rings), it can position itself between base-pairs, in

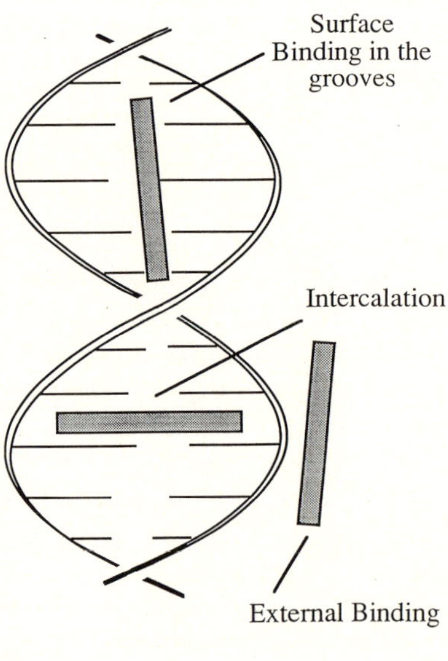

Fig. 5. Schematic representation of the surface binding, intercalation, and external binding in the atmosphere of ions around the DNA for small molecules

a sandwich-like complex. This phenomenon, called intercalation [30, 31], was first described by Lerman [32] to explain the binding of aminoacridines to DNA. Intercalation increases the separation of adjacent base pairs (3.4 Å) and the resulting helix distortion is compensated by the adjustments in the sugar-phosphate backbone and generally by an unwinding of the duplex (26° for ethidium bromide, Fig. 6). Association constants for usual intercalators are of the order of 10^5 to 10^6 M^{-1} [30].

Surface binding (Fig. 5) takes place in the minor or major groove, although minor groove binding is more common for small molecules. The electrostatic potential, the pattern of hydrogen accepting or donating groups, and the narrowness of the minor groove are three characteristic factors which are different for GC and AT base pairs, and are responsible for a preferential binding of organic molecules in AT tracts. The presence in the minor groove of the N-2 amine group on guanine (in GC base pairs) is also an important factor orienting the site of binding.

The external binding mode (Fig. 5) is due mostly to the electrostatic interaction of cations with the negatively charged phosphate backbone at the periphery of the double helix [33].

We illustrate below the relative importance of the different types of interactions for surface binding in minor grooves with the example of netropsin (Fig. 6). The binding, preferentially at AT base pairs [23, 34–37], of this antiviral and antibiotic involves both an electrostatic component, from the two cationic ends, and the formation of hydrogen bonds with the central three amide NH groups. The specificity in DNA-binding (A vs G) arises in fact from steric hindrances between the NH_2 group of the guanine and two types of netropsin atoms [35]: either the CH of the pyrrole ring, or the CH_2 flanking the outer amides on the wings of the drug molecule (Fig. 6). Thus, when binding to double-helical DNA,

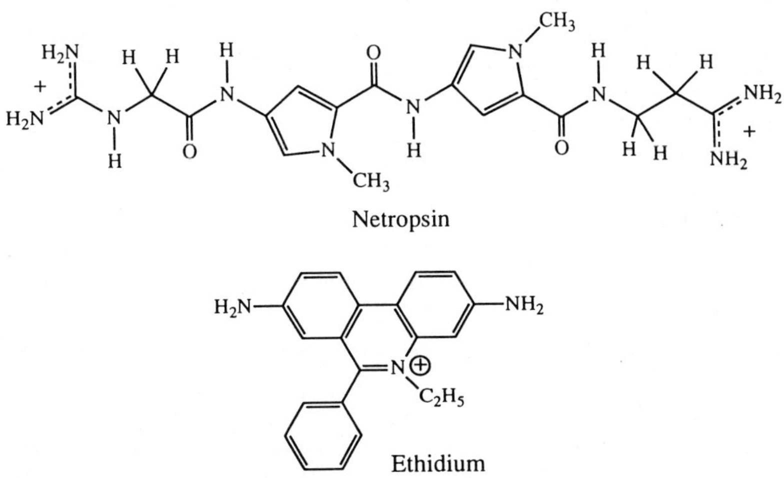

Fig. 6. Structures of ethidium bromide and netropsin

the drug displaces the spine of hydration surrounding the DNA, which helps to stabilize the B-form DNA, and replaces it by the backbone of the drug molecule. This removal of water molecules contributes to the entropy increase of the DNA-drug binding, while the electrostatic, H-bonding, and London interactions probably contribute to the enthalpy of binding.

3.3 A Key Question – Does the Molecule Intercalate or Surface Bind?

The answer can be given by analysing the system DNA-molecule by a battery of complementary methods [38, 39] (Table 1, first column), keeping in mind that the binding mode is also influenced by the composition of the solution.

Crystallographic determination of the "complex" formed between the interacting molecule and DNA or oligonucleotides, constitutes obviously the method yielding the most accurate picture of the mode of interaction with nucleic acids in the solid state. Thus intercalation or surface binding in oligonucleotides major or minor grooves have been demonstrated for several organic molecules by X-ray spectroscopy [23, 31]. For metallic complexes however, there exist only a few cases where structures have been characterised by X-ray. We can mention the example of the Pt complexes, generally bound covalently [40] or by intercalation of a polypyridyl ligand [41], and the case of $Ru(NH_3)_6^{3+}$ bound by hydrogen-bonding within the nucleic acid grooves [42]. Obviously the drawback of such crystallographic methods is the fact that, in the solid state, the geometry of these macromolecules may be different than that in solution or in a cell, and indeed one particular molecule may bind by intercalation or surface binding depending on the conditions in the solution.

In the absence of crystallographic data, there exist many other experimental criteria that allow the determination of the binding mode [30, 38]. They can be classified into two categories: the measurement of physical effects on DNA and spectroscopic studies (Table 1, first column, C and A respectively). Irrespective of the binding mode, the binding parameters (the affinity constant and the number "n" of occupied base pairs per molecule) can be determined from equilibrium dialysis and Scatchard plots [43–46]

3.3.1 Physical Effects on DNA

The methods giving access to the effects of interacting molecules on the physical properties of DNA are cited first because they are considered as the most classical ones for testing intercalation, although in the particular case of the metallic complexes developed below, they have been generally less applied than the spectroscopic techniques.

Among these methods, hydrodynamic measurements of sedimentation coefficients and viscosity are classical [43, 47]. The insertion of a true intercalator (ethidium bromide for example) between the stacking of bases produces

Table 1. Study of interaction of $Ru(Phen)_3^{2+}$ with DNA, methods and observations

Methods	Observations	Consistent (+), or inconsistent (−) with intercalation	Ref.
A. Spectroscopy			
1. Absorption, luminescence	– Hypochromicity in absorption, enhanced luminescene when + B-DNA, more important for Δ than for Λ	+	48
2. Emission lifetimes (τ)	– Increase of τ when + B-DNA, same for Δ and for Λ	+ −	48
3. Differential quenching in emission intensity or lifetime	– Quenching by $Fe(CN)_6^{4-}$ suggests the presence of 2 species on DNA: intercalated + surface bound	+	51, 52
4. Steady-state emission polarization (P)	– Rac. + B-DNA: increase of P;	+	51
	– B-form poly[d(G-C)]: P increases more with Δ	+	52
	– Z-form poly[d(G-C)]: P increases more with Λ	+	70
5. Linear dicroïsm (LD), orientation by electric field	– Conclusion from estimated angles between transition dipole and helix axis: Δ intercalates into B-DNA, not Λ	+	53, 54
LD, orientation by flow field	– Conclusion from estimated angles between the 3-fold axis of Δ and B-DNA helix axis: not in accordance with intercalation of Δ	−	55
	– Δ decreases flow of B-DNA orientation, (reverse expected for intercalation)	−	55
	– LD spectra of Δ + B-DNA as a function of ionic strength and binding ratios suggest one binding mode or 2 with the same electrostatic contribution to the binding energy	−	55, 56
6. Equilibrium dialysis + CD	– When rac. + B-DNA or Z-DNA, binding of Δ is favoured and induces Z to B transformation	+ −	57
	– When rac. + B-DNA, binding of Δ is favoured, when rac. + Z-DNA (very high ionic strength), no optical enrichment, for low binding levels no Z to B transformation	+	48
7. ^{1}H-NMR two dimensions, NOE	– With $d(CGCGATCGCG)_2$, both Δ and Λ bind into minor groove of the central AT-TA region; rapid exchange between the bound and free complex in the NMR time scale	−	60, 61
	– $d(GTGCAC)_2$ and 5′-$pd(CGCGCG)_2$ act as chiral shift reagents for the complex. More surface binding into minor groove than intercalation	+ −	62, 63
8. EPR	– With complex derivatized by stable nitroxide: detection of 3 different degrees of binding	+	64

Table 1. (contiued)

Methods	Observations	Consistent (+), or inconsistent (−) with intercalation	Ref.
B. Molecular modeling			
9. Energy minimization calculation	– B-DNA intercalated and surface-bound species exhibit more or less the same binding energies, probable steric effects for Δ and Λ in minor grooves	+	84
C. Physical DNA properties			
10. Relative viscosity	– Δ + B-DNA decreases relative viscosity, thus no classical intercalation	–	43
	– Λ + B-DNA has almost no effect, thus Λ surface binds in the grooves	–	43
11. DNA unwinding			
– Topoisomerisation	– unwinding of supercoiled DNA of 22° per bound complex	+	49
– Electrophoretic migration	– Δ more than Λ reversibly unwinds supercoiled DNA	+	48
12. Thermal denaturation	– with poly[d(A-T)] + complex, behaviour similar to that with ethidium bromide.	+	49

Abbreviations: Δ: enantiomer Δ; Λ: enantiomer Λ; rac: racemic; + —: indicates that the observation is not in favor nor against intercalation; LD: linear dichroïsm; CD: circular dichroïsm a lengthening of the DNA double helix and consequently an increase of viscosity, whereas a pure surface bound species (such as antibiotic Hoechst 33258) does not affect the viscosity at all [43]. Due to this lengthening, DNA unwinding is also produced, i.e. a drop of the angle between the long axis of succesive basepairs; typical intercalators unwind DNA between 17° and 26° per bound species [30]. A molecule can produce DNA unwinding without lengthening, in which case it is probably not a true intercalator (see further, for $Ru(phen)_3^{2+}$, Fig. 2). The unwinding of supercoiled DNA can be measured by its migration through 1% agarose [48], or by topoisomerisation experiments [49, 50].

Finally, thermal denaturation studies of double stranded nucleic acids in the presence of the interacting species allow one also to draw conclusions on the intercalating abilities of a molecule [49] as the behaviour observed with classical intercalators and pure electrostatic binders are quite different.

3.3.2 Spectroscopic Methods

UV-visible absorption and emission spectroscopy. Different pieces of information can be obtained from this spectroscopy according to the type of method which is used. First indications of intercalation may appear in the absorption and emission spectra of the binding species. With intercalators the π interaction with the stacked bases induces hypochromic effects and shifts of the absorption band to longer wavelengths. The emission intensity may be enhanced with intercalation [48] and in parallel with these spectroscopic changes, the luminescence lifetimes are increased. Sometimes several lifetimes may be detected, for example a short one with a value corresponding to the emitting species in aqueous solution in the absence of DNA, and a longer one; that means that the emitting species are distributed partially on the DNA and in solution [51]. Differential luminescence quenching [51, 52] allows one to refine the binding picture and thus to differentiate well-protected DNA species from less well-protected ones, from molecules in solution.

Steady state emission polarization measurements reflect the time over which the excited species remains rigidly bound to DNA, yielding some idea of the movement and orientation of the luminophore in the DNA microenvironment.

Linear dichroism data with DNA oriented by an electric field [53, 54] or a linear flow [55, 56], under linearly polarised light, lead to determinations of the angle between the absorbing transition dipole moment of the chromophore in the molecule and the DNA helix axis; conclusions concerning intercalation may thus be drawn from this technique. Finally, with chiral compounds, circular dichroism is also an attractive method to determine the enantioselectivity in the binding of the molecule [48, 57, 58].

NMR spectroscopy. NMR is a powerful tool for the analysis of the structure and dynamics of drug-nucleic acid complexes [59], and has been widely used for characterising the binding modes of organic molecules with oligonucleotides. It has been less applied so far for metal complexes [60–63].

EPR spectroscopy. Molecules can also be derivatised by stable nitroxides used as spin probes [64] which allows one to draw conclusions on the microenvironment of the interacting molecule, thus on its binding.

In conclusion, each technique yields different kinds of information, such as the number of different microenvironments surrounding the molecule, the tightness of the interaction or restriction of the compound movement, the geometry or the orientation of the ligands versus the base pairs. A factor which plays an important role and which should not be forgotten when comparing the results from the different techniques, is the different time scales associated with each of these methods. For example, in the NMR time scale, the bound and free molecule (if it exists) exchanges rapidly, whereas in the time scale of the emission lifetimes, one can differentiate the species emitting from the DNA from those emitting from the solution.

3.4 Binding of Ru(II) Polypyridyl Complexes to DNA

Below, we consider some of the binding studies reported for the metal complexes. We present successively examples of complexes interacting by surface binding and intercalation. We finish with the controversial case of $Ru(phen)_3^{2+}$ whose studies have triggered the research on interactions and photoreactions of Ru(II) polypyridyl complexes with DNA. This complex illustrates the problems which can arise when comparing the results from the different techniques described above. One should also always remember that for metal complexes, binding modes may be more complicated than for organic molecules, and that neither intercalation, nor groove binding are unambiguous concepts but rather names used to denominate a group of DNA binding modes having important common features.

For an exhaustive list of complexes whose interaction with DNA has been examined, the reader should refer to Table 2; several compounds are also described in [65].

3.4.1 Surface Binding

One example of complex showing a typical surface binding, is $Ru(TMP)_3^{2+}$ (TMP = 3,4,7,8-tetramethyl phenanthroline, Fig. 2) [66, 67] which does not intercalate at all because of the methyl groups. This complex turned out to be a shape-selective probe for A-DNA conformation (see also Sect. 5), where it fits well into the shallow and wide minor-groove surfaces. This surface-bound species is apparently free to diffuse along the helix surface and is thus less rigidly bound than an intercalated complex. $Ru(bpy)_3^{2+}$ may also be considered as a complex which does not intercalate as it does not unwind DNA [49]. The luminescence enhancement found for $Ru(bpy)_3^{2+}$ upon binding to DNA shows strikingly greater sensitivity to increased ionic strength [49, 68] than $Ru(phen)_3^{2+}$ (see below), confirming a different type of binding mode. While it is

Table 2. List of polypyridyl Ru(II) complexes studied in the presence of DNA

Complexes	Interactions	Photoreactions	Miscellaneous
$Ru(phen)_3^{2+}$	43, 44, 48, 49, 51, 53, 54 55, 56, 60, 61, 62, 68, 70, 73, 74, 77, 81, 95 84, 141, 191, 192, 193	49, 74, 97, 107, 117, 119,141	77, 81, 113, 114, 119, 192,
$Ru(bpy)_3^{2+}$	49, 68, 74, 96, 109, 54, 60, 70, 193	49, 74, 96, 97, 117, 105, 106, 107, 108	113, 195
$Ru(bpy)_2phen^{2+}$	73, 74, 193	74	
$Ru(bpy)_2DIP^{2+}$	73, 74, 193	74	
$Ru(bpy)_2biq^{2+}$	74	74	
$Ru(bpy)_2phi^{2+}$	193, 194		
$Ru(bpy)_2HAT^{2+}$	73, 74, 75, 95	74	
$Ru(bpy)_2DPPZ^{2+}$	7, 69, 70		7
$Ru(bpy)_2PPZ^{2+}$	77, 79, 80		77, 78
$Ru(bpy)_2TAP^{2+}$	73, 96	73, 96	
$Ru(bpy)_2dpp^{2+}$	77, 80		
$Ru(bpy)_2qpy^{2+}$	80		
$Ru(bpy)_2Meqpy^{3+}$	80		
$Ru(bpy)_2Me_2qpy^{4+}$	80		
$Ru(phen)_2bpy^{2+}$	193		
$Ru(phen)_2DIP^{2+}$	193		
$Ru(phen)_2flone^{2+}$	193		
$Ru(phen)_2DPPZ^{2+}$	72		
$Ru(phen)_2phi^{2+}$	193		
$Ru(phen)_2en^{2+}$	74	74	
$Ru(phen)_2HAT^{2+}$	73, 95		
$Ru(HAT)_3^{2+}$	73	73	
$Ru(TAP)_2HAT^{2+}$	73, 81, 95	73	81
$Ru(HAT)_2TAP^{2+}$	73	73	
$Ru(TAP)_3^{2+}$	74, 95, 96, 97, 119	74, 96, 97, 100, 119, 124	100
$Ru(TAP)_2bpy^{2+}$	73, 96	96	
$Ru(bpy)(HAT)_2^{2+}$	73	73	
$Ru(bpy)(TAP)(HAT)^{2+}$	73	73	
$Ru(DIP)_2phen^{2+}$	193		
$Ru(DIP)_3^{2+}$	70, 192, 196		192
$Ru(biq)_2bpy^{2+}$	74	74	
$Ru(5\text{-}NO_2\text{-}phen)_3^{2+}$	193		
$Ru(TMP)_3^{2+}$	66, 67, 73, 141,	67, 141, 66	
$Ru(phi)_2bpy^{2+}$	193		
$Ru(DMB)_2phen^{2+}$	74	74	
$Ru(tpy)_2^{2+}$	74, 49	74	
$Ru(phen)_2CN_2$	74, 49	74	
$Ru(bpy)_2CN_2$	49		
$Ru(DIP)_2Macro^{n+}$			197, 198
$Ru(TAP)_2POQ^{2+}$	211		
$Ru(bpy)_2(phen\text{-}T)^{2+}$	64		
$Ru(phen)_2(phen\text{-}T)^{2+}$	64		
$Co(phen)_3^{3+}$	62, 200, 208	117, 147	
$Co(bpy)_3^{3+}$	200	117	
$Co(en)_3^{3+}$		117	
$Co(NH_3)_6^{3+}$		117	
$Co(DIP)_3^{3+}$	148, 150, 192	147, 148, 150, 192	150, 192
$Cr(phen)_3^{3+}$	63	117	
$Ni(phen)_3^{3+}$	63		
$Zn(phen)Cl_2$	206		
$Zn(phen)_2^{2+}$	206		

Table 2. (contiued)

Complexes	Interactions	Photoreactions	Miscellaneous
$Zn(phen)_3^{2+}$	206		
$Os(phen)_3^{2+}$	73		
$Fe(phen)_3^{2+}$	44, 54, 58, 208		
$Fe(bpy)_3^{2+}$	208, 210		
$Ru(bpy)_3^{3+}$			177
$Rh(phen)_2phi^{3+}$	136, 139, 140, 141, 201	136, 139, 140, 141	
$Rh(phen)_3^{3+}$	62	117	
$Rh(phi)_2bpy^{3+}$	139, 136	136, 139, 202	
$Rh(DIP)_3^{3+}$	141, 138	138, 141	
$Rh(phen)_2Cl_2^{1+}$	142	142, 143	
$Ru(tpy)bpy(OH_2)^{2+}$	205		203, 204, 205
$Ru(tpy)phen(OH_2)^{2+}$	205		204, 205
$Ru(bpy)_2py(OH_2)^{2+}$			204, 209
$Ru(phen)_2(OH_2)_2^{2+}$			204
$Ru(tpy)tmen(OH_2)^{2+}$	205		204, 205
$Ru(DPPZ)tpy(OH_2)^{2+}$	207		207
$Ru(phen)_2py(OH_2)^{2+}$			204, 209
$Ru(tpy)tpt(OH_2)^{2+}$			209
$Ru(tpy)tmen\text{-}AO(OH_2)^{2+1}$			209
$Ru(bpy)_2(OH_2)_2^{2+}$			204
$Ru(phen)_2Cl_2$			199

possible that $Ru(bpy)_3^{2+}$ is merely loosely positioned in the ion atmosphere of the DNA polyelectrolyte, it may also be held in the groove by the electrostatic potential.

3.4.2 Intercalation

With an appropriate heteroaromatic ligand, showing extended aromaticity, combined with properly chosen ancillary ligands, the resulting metallic compound may behave as an intercalating agent. In this regard, as an octahedral complex is present in the form of Λ and Δ enantiomers (Fig. 7) that can be separated, each enantiomer can be studied separately. Thus intercalation, where one of the ligands of the complex is sandwiched between the nucleo-bases (Fig. 8), has been demonstrated for a few complexes, for the Λ and Δ enantiomer versus a right-handed helix (or B-DNA). A complete intercalation of this ligand is of course prevented because of the steric hindrance of the two non intercalated ancillary ligands.

Such pictures correspond to Ru(II) complexes designed with the DPPZ ligand (dipyrido [3,2-a:2′3′-c] phenazine, Fig. 2) [69–72], that have been studied after the $Ru(phen)_3^{2+}$. The shape and aromaticity of DPPZ allows indeed a good insertion between the stacked bases. $Ru(bpy)_2DPPZ^{2+}$ [69, 70] (enantiomeric mixture) has a binding constant for B-DNA much higher than the other complexes and even higher than ethidium bromide ($K > 10^6\ M^{-1}$). Results from topoisomerisation experiments yield an unwinding angle of 30° consistent with intercalation. This complex does not luminesce at all in aqueous solution

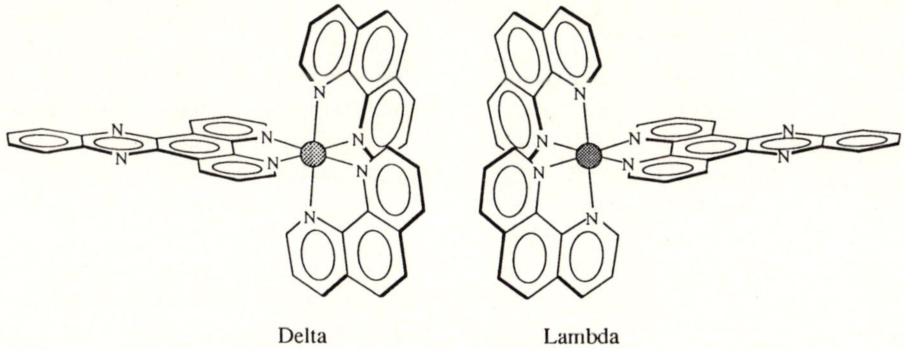

Fig. 7. The Δ and Λ enantiomers of $Ru(phen)_2DPPZ^{2+}$

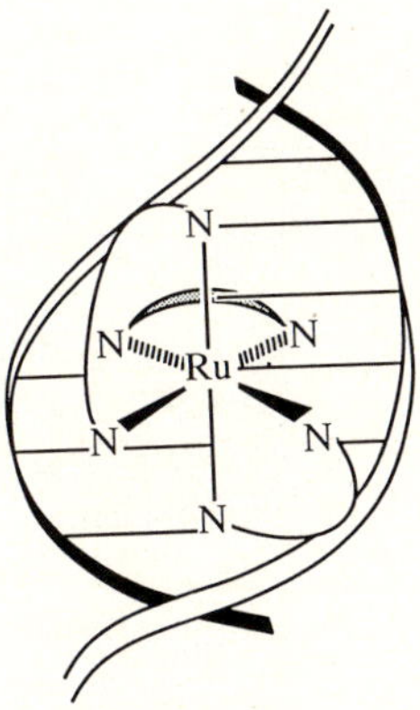

Fig. 8. Schematic representation of intercalation into a DNA groove for an octahedral trischelated complex [adapted from Barton JK, Danishefsky AT, Goldberg JM (1984) J. Am. Chem. Soc. 106: 2172]

and its luminescence is switched on by interaction with DNA because of the protection of the DPPZ ligand from the aqueous environment. The two luminescence lifetimes observed with $Ru(bpy)_2(DPPZ)^{2+}$ and derivatives, in the presence of B-DNA, have been attributed to complexes with different intercalation geometries [71]. From polarisation measurements, an intercalation association is also proposed for $Ru(bpy)_2DPPZ^{2+}$ (enantiomeric mixture) with Z-DNA [70]. A detailed binding study has been carried out with the separated enantiomers of $Ru(phen)_2DPPZ^{2+}$ [72] (Fig. 7).

Complexes with less extended aromaticity such as $Ru(bpy/phen)_2HAT^{2+}$ [73–76] (HAT = 1,4,5,8,9,12-hexaazatriphenylene, Fig. 2) and $Ru(bpy)_2PPZ^{2+}$ [77–80] (PPZ = 4,7-phenanthrolino-[6,5-b] pyrazine, Fig. 2) exhibit also characteristics most relevant to intercalation. We can mention: (1) a very slow mobility of the HAT complex along the DNA double helix [81], (2) a good protection of the complex versus reagents that remain in the bulk solution [73, 79], and (3) a clear hypochromic effect on the MLCT transition in the presence of DNA [73, 75, 79, 80].

3.4.3 The Controversial Case of Ru(phen)$_3^{2+}$

The values of the binding constants determined with different salt concentrations by equilibrium dialyses [43, 48], luminescence titrations and electrochemiluminescence [82], are all 2 or 3 orders of magnitude lower than for ethidium bromide. Therefore, a priori, they do not indicate contribution of "classical intercalation" into DNA as described for organic molecules and for the DPPZ, HAT and PPZ complexes.

The results and observations from the experimental methods used to study the interaction modes of Ru(phen)$_3^{2+}$ are compiled in Table 1. The examination of this table indicates obvious disagreements between the authors concerning the intercalation of Ru(phen)$_3^{2+}$ into DNA. Chronologically, the first spectroscopic experiments (entries 1 to 4) and the first results on DNA unwinding and denaturation (entries 11, 12) in 1984–1986 were all consistent with intercalation. Afterwards, with the results from LD and NMR in 1988–1992 (entries 5, 7) and with the viscosity measurements in 1992 (entry 10), the intercalation of Ru(phen)$_3^{2+}$ has become questionable.

Thus on the one hand, from a series of approaches (entries 1–4, 6–8, 11), Barton and co-workers describe 3 modes of binding for this complex, some of them with enantiomeric selectivity:

(1) a "partial intercalation" generally from a major groove, with some enantiomeric preference between species with the same helicity. Thus the Δ enantiomer binds preferentially to a B-DNA (right-handed helix), and vice versa some data also support intercalation of the Λ enantiomer into a left-handed DNA or Z-DNA (entry 4, [70]).

(2) a surface binding along the DNA grooves, which was first suggested to take place in the major grooves [51], and afterwards in the minor grooves, based on the NMR results with oligonucleotides [62, 63]. With B-DNA there is a weak preference for the Λ enantiomer, showing that for this mode of binding it is the complementarity of helicity which controls the slight enantiomeric selectivity.

(3) an unbound complex is not "totally free" but probably loosely associated with the polyanionic DNA in the atmosphere of ions, insensitive to DNA handedness or groove size. This third mode of interaction (as well as the others), has in fact been detected with a nitroxide derivative of Ru(phen)$_3^{2+}$ by EPR (entry 8).

On the other hand, these conclusions seem difficult to reconcile, a priori, with those from linear dichroism (entry 5), NMR spectroscopy (entry 7), and viscosity measurements (entry 10), which strongly suggest a surface binding. A new mode of interaction should thus be proposed which fits all the results.

Satyanarayana et al. [43] (entry 10) have made interesting suggestions in this direction. Based on the study of the binding constant and on the fact that the relative calf-thymus-DNA viscosity decreases by the addition of Δ Ru(phen)$_3^{2+}$, it is concluded that, if some kind of intercalation is present, it is certainly not a classical one. The model proposed many years ago by Kapicak and Gabbay

[83] to explain a decrease of viscosity could be adopted for the Δ complex. According to this model, the partial intercalation of the molecule results into a static bend (or kink) of the DNA helix, which would reduce its effective length and concomitantly its viscosity. As the Λ enantiomer has almost no effect on the viscosity (entry 10), it is concluded that it surface binds, in accordance with other results gathered in Table 1.

Such a kink induced by a partial intercalation of the Δ complex does not exclude the fact that the complex could unwind the helix (entry 11) without lengthening [43]. In other words the unwinding helix can be regarded as a necessary, but insufficient criterion for "classical" intercalation. This picture would also be consistent with the fact that the Δ enantiomer decreases the flow of B-DNA orientation in LD study (entry 5), and that the LD spectra would suggest two modes of binding (for example surface binding + partial intercalation with a kink) with the same electrostatic contribution to the binding energy, as also indicated from molecular modeling [84] (entry 9). Obviously with this picture, unambiguous interpretation of LD data concerning the calculation of the angle of orientation is difficult. This model of interaction is also compatible with the conclusions from luminescence lifetimes measurements and EPR spectroscopy. It is indeed reasonable to conceive that the microenvironments "felt" by the "partially intercalated" Δ complex leading to a DNA kink, would be different from the one probed by the surface-bound Λ isomer, these latter two are also different from the ionic atmosphere around the helix.

A derivative of $Ru(phen)_3^{2+}$, $Ru(DIP)_3^{2+}$ (DIP = 4,7-diphenyl 1,10-phenanthroline, Fig. 2) behaves similarly to $Ru(phen)_3^{2+}$ [65, 85] with a slightly better enantioselectivity, so that the Λ $Ru(DIP)_3^{2+}$ has been used as spectroscopic probe for Z-DNA. This enantioselective property has been applied with photoreactive complexes, where the Ru(II) center has been replaced by another metal ion such as Co(III) and Rh(III) (see further).

4 Photophysics and Photochemistry of Ru(II) Polypyridyl Complexes

4.1 Behaviour in the Absence of DNA

The photophysics and photochemistry of Ru(II) complexes have been extensively studied and good reviews are available on this subject [1, 86, 87]. The type of reactivity associated with Ru(II) polypyridyl complexes in the excited state depends on the nature of this excited state and consequently on the different possible photophysical pathways controlling the luminescence lifetimes. For most polypyridyl Ru(II) complexes (for example $Ru(bpy)_3^{2+}$, $Ru(phen)_3^{2+}$, $Ru(bpz)_3^{2+}$ ... Fig. 2) [1, 86, 87], population of the excited singlet MLCT state is followed by crossing to the triplet MLCT state (^{3}MLCT) with a quantum yield of 1 [1, 86]. From this triplet, the ^{3}MC state (Metal Centered triplet) can be reached by thermal activation, which is of course dependent on the energy difference between the two states (^{3}MLCT and ^{3}MC) [1, 88, 89] (Fig. 9). Two different reactivities are associated with the two states. In the ^{3}MLCT state, the complex generally gives rise to redox processes, whereas in the ^{3}MC state, the complex being distorted, the Ru-nitrogen bond weakens so that a rupture of a single Ru–N bond can take place [90] (Fig. 9). If this latter process is followed by a second Ru-nitrogen break, there is a ligand loss, which we will call photodechelation. However, depending on the flexibility of the ligand (with bpy for example), the first bond break can give rise to an intermediate compound where the leaving ligand has retained one Ru-nitrogen bond and where the empty coordination site has been filled by a monodentate ligand such as a water molecule or a chloride ion. This intermediate can regenerate the starting material or give rise to a ligand loss.

Whether a photoredox or a photodechelation process takes place, depends on the presence of reducing or oxidising agents in the solution, and on the energy difference between the ^{3}MLCT and ^{3}MC. An estimation of this energy

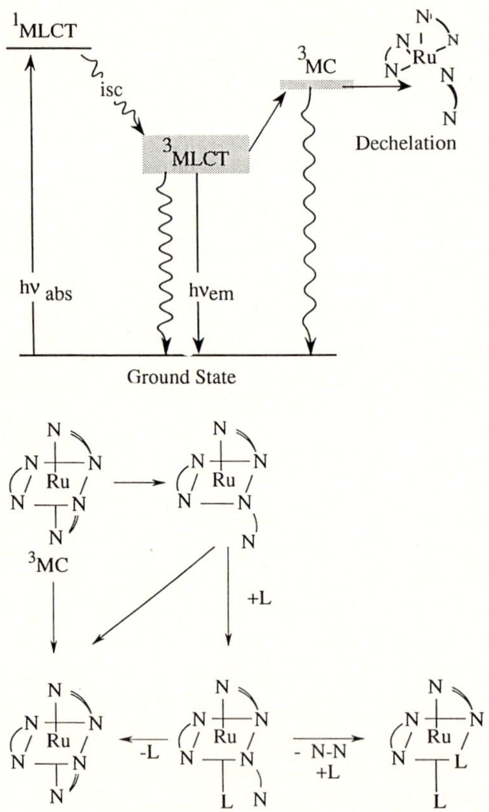

Fig. 9. Diagram of the excited states levels of Ru(II) polypyridyl complexes with different pathways for the photodechelation [adapted from Durham B, Casper JJ, Nagle JK, Meyer TJ (1982) J. Am. Chem. Soc. 104: 4803]

can be obtained from the analysis of the dependence of the luminescence lifetimes on temperature, for a series of complexes containing the same charge transfer (CT) excited state [88, 90, 91]. This is illustrated in Fig. 10 for complexes based on the $[Ru^{3+}\text{-}HAT^{\bullet -}]^*$ charge transfer state [91]. The energy of the ^{3}MLCT decreases with the decreasing number of HAT ligands in the compound, whereas the ^{3}MC energy remains almost constant. Moreover, as indicated by the excited state reduction potentials in the same figure, the oxidising power associated with the ^{3}MLCT state increases with the increasing number of HAT ligands in the complex. The consequences of these redox and energetic considerations are important [91]. For $Ru(bpy)_2(HAT)^{2+}$ the ^{3}MC is not populated at room temperature; thus only photoredox processes with the ^{3}MLCT take place, and for photoreduction, only with relatively strong reducing agents. For $Ru(HAT)_3^{2+}$, as both excited states are populated, both redox and dechelation processes are observed. Therefore, $Ru(HAT)_3^{2+}$ under illumination leads not only to dechelation, but is also capable of abstracting electrons from donors which are rather poor reductants. The relative contributions of these two processes depends on the concentration and reducing power of the reductant. A reductive quenching of the ^{3}MLCT should compete efficiently with the conversion of the ^{3}MLCT into the ^{3}MC.

The trends in the properties described for this series of HAT complexes, are similar for the corresponding series based on the TAP ligand [92], although HAT complexes are more oxidising than TAP complexes. As will be illustrated further, the modulation of the oxidation power will be reflected in the photoreactivity of the Ru(II) complexes versus the various DNA bases.

It is also noteworthy that complexes containing ligands such as TAP, HAT, bpz (2,2′-bipyrazine) or bipym (2,2′-bipyrimidine) (Fig. 2), have free non-chelated nitrogen atoms. It has been shown that all these compounds are all more basic in the excited state than the ground state [75, 93, 94], so that the excited states are already protonated at pH 5–6 on the non-chelated nitrogen atoms.

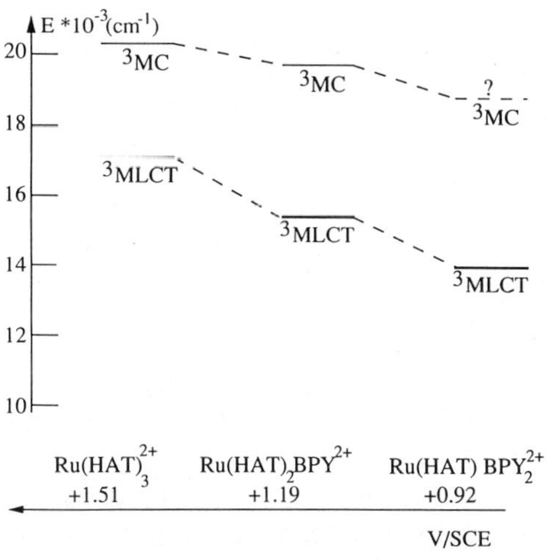

Fig. 10. Energies of the ^{3}MLCT and ^{3}MC states in the series of HAT complexes and corresponding E^0 (Ru^{2+*}/Ru^{1+}) [adapted from: Jacquet L, Kirsch-De Mesmaeker A (1992) J. Chem. Soc. Faraday Trans. 88: 2471]

4.2 Photophysics in the Presence of DNA and Mononucleotides

As mentioned before in Sects. 3.3. and 3.4., the luminescence intensity and lifetimes are often enhanced when the complexes interact with DNA. This has been observed with complexes that intercalate (Sect. 3.4, *intercalation*) and also with $Ru(phen)_3^{2+}$ and derivatives. This emission increase originates from a protection of the complex by the DNA versus O_2 quenching [74] and from the effect of the DNA environment on the non-radiative deactivation rate constants controlling the ^{3}MLCT lifetime. The relative importance of these two effects depends on the efficiency of the O_2 luminescence quenching and on the photophysical mechanism controlling the luminescence. For complexes whose luminescence in water is insensitive to O_2 [for example for $Ru(bpy)_2HAT^{2+}$ (Fig. 10)], due to the large energy gap between the MLCT-MC states, it is mainly the non-radiative deactivation rate constant (k_{nr}) associated with the ^{3}MLCT state which controls the room temperature luminescence lifetime. Therefore the influence of DNA can be detected only in the value of this rate constant [95]. In fact k_{nr} decreases in the presence of DNA (leading to a lifetime increase), probably due to some protection of the complex from the aqueous phase by the hydrophobic double helix environment. This is in sharp contrast to the case of $Ru(TAP)_3^{2+}$ and $Ru(TAP)_2(HAT)^{2+}$ where the emission lifetime which is unaffected by the presence of O_2 in water, is mainly governed by the thermal activation rate constant towards the ^{3}MC state (k_{MC}) [91]. Thus it is concluded, from the photophysical results, for those complexes that the nucleic acid (poly[d(A-T)] in this case) microenvironment decreases both k_{MC} and k_{nr} [95], leading to an enhanced luminescence lifetime. It is probable that in this case the emprisonment of the complex within the double helix, responsible for a drop of k_{MC}, would prevent in some way the distortion of the compound in the ^{3}MC state, leading to a less efficient loss of ligand [95] [124].

For $Ru(bpy)_2(HAT)^{2+}$, [i.e. the least oxidising compound in the series of HAT complexes (Fig. 10)], the luminescence enhancement, in intensity and lifetime, is observed in the presence of poly[d(A-T)] as well as with CT-DNA and poly[d(G-C)] [73–75] (case 1). However with $Ru(bpy)(HAT)_2^{2+}$, this luminescence enhancement is observed only in the presence of poly[d(A-T)] [73]; in contrast, a luminescence quenching occurs with increasing amount of CT-DNA or poly[d(G-C)], i.e. the most easily oxidisable polynucleotides (case 2). Moreover with $Ru(HAT)_3^{2+}$ the emission is not only quenched by CT-DNA, but also by poly[d(A-T)] (case 3) [73]. Figure 11 illustrates the case 2 behaviour of another complex, $Ru(TAP)_3^{2+}$. The three cases are also observed for the other complexes of the series $Ru(TAP/HAT)_{3-n}(bpy/phen)_n^{2+}$ ($n = 0, 1, 2$) [96, 97]. Consequently the quenching by the polynucleotides follows the increasing oxidation power of the excited complex, as the most powerful oxidising complexes are able to photo-oxidise poly[d(A-T)], i.e. the less easily oxidisable polynucleotides. When there is an important emission quenching under steady state illumination, the analysis of the luminescence decays under pulsed illumination (single photon counting in the ns timescale)

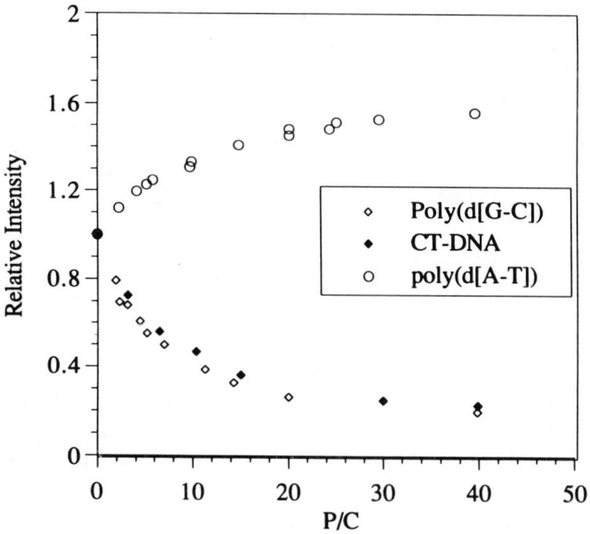

Fig. 11. Relative emission intensities of $Ru(TAP)_3^{2+}$ as a function of the ratio DNA phosphate/complex P/C) in air saturated phosphate buffer (3 mM) ($\lambda_{em} = 600$ nm; $\lambda_{exc.} = 450$ nm) [adapted from: Kelly JM, Mc Connell D, OhUigin C, Tossi AB, Kirsch-De Mesmaeker A, Masschelein A, Nasielski J (1987) J. Chem. Soc., Chem. Commun. 1821]

leads to monoexponential or quasi monoexponential functions, with lifetimes corresponding to those of the complexes in pure aqueous solution. These observations indicate the existence of a static quenching of the excited species interacting within DNA.

Emission quenching is also observed with mononucleotides. In that case the quenching efficiency decreases from GMP (guanosine 5′ monophosphate) to AMP (adenosine 5′ monophosphate) i.e. it also follows the redox potentials of the bases, as G is more easily oxidisable than A, although the oxidation potential values reported in the literature are rather different from one author to the other [101–104]. Moreover the quenching rate constant by GMP in a series of different TAP and HAT complexes plotted versus the reduction potential of the excited state (Fig. 12) [95] is consistent with an electron transfer process. Indeed, as will be demonstrated in Sect. 4.3.1, these quenchings (by the mono- and polynucleotides) originate from such processes.

4.3 Photo-Electron Transfer Processes in the Presence of Mononucleotides and DNA

In this Section, two types of photo-electron transfer processes with the ^{3}MLCT state of complexes will be successively discussed. We will first introduce the direct photo-electron transfer from a DNA base to the excited complex Sect. 4.3.1. Afterwards we will consider the electron transfer between an excited

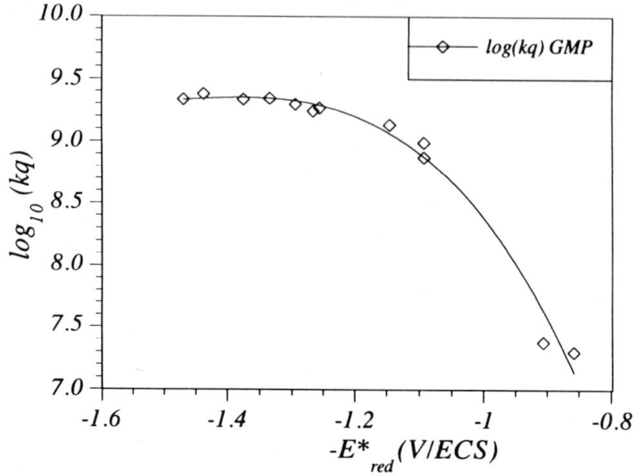

Fig. 12. Rate constants of luminescence quenching by GMP for a series of complexes as a function of the reduction potential of their ^{3}MLCT states. [adapted from: Lecomte J-P. (1992) Ph.D. thesis, Brussels, Belgium]; From the left to the right: $Ru(HAT)_3^{2+}$, $Ru(HAT)_2TAP^{2+}$, $Ru(TAP)_2HAT^{2+}$, $Ru(TAP)_3^{2+}$, $Ru(bpz)_3^{2+}$, $Ru(diCH_3TAP)_3^{2+}$, $Ru(HAT)_2phen^{2+}$, $Ru(HAT)_2bpy^{2+}$, $Ru(TAP)_2bpy^{2+}$, $Ru(phen)_2HAT^{2+}$, $Ru(bpy)_2HAT^{2+}$

complex and an additive, oxidant or reductant, either remaining in solution or bound to DNA, Sect. 4.3.2. The reactions of DNA resulting from these primary processes will be described in Sect. 5. For an exhaustive list of Ru(II) complexes whose photoreactivity with DNA has been examined, the reader should refer to Table 2.

4.3.1 Photo-Electron Transfer from a DNA Base to the Excited Complex

Direct photo-electron transfer from a DNA base to excited Ruthenium(II) complexes has been observed with complexes with two or three chelated TAP, HAT or bpz ligands, i.e. with very strongly oxidising complexes. Their strong MLCT absorption in the visible region, and strong emission at room temperature, with lifetimes often longer than 100 ns [1, 86, 91] make a study of their photoreactivity with DNA rather easy. Moreover the nature of the photoreactive state (MLCT versus MC) (see Sect. 4.1), the reduction potentials of the excited states, as well as the spectra of reduced complexes, have been reported in the literature [86, 91, 98, 99]. Such good knowledge of the properties of the complexes under illumination is an essential step for the study of their photoreactions with DNA.

The photoreactions of $Ru(TAP)_3^{2+}$ in the presence of different mononucleotides have been used as models for the photoreactivity with CT-DNA and polynucleotides and for the photoreactions of other oxidising polyazaaromatic Ru(II) complexes [100]. Flash photolysis experiments with $Ru(TAP)_3^{2+}$ in the presence of GMP or AMP, demonstrate clearly the presence of a photo-induced electron transfer from GMP, and to a lesser extent from AMP, to the excited complex [100].

$$Ru(TAP)_3^{2+*} + GMP \longrightarrow [Ru(TAP)_2(TAP^{\bullet -})]^+ + GMP^{\bullet +} \quad (14)$$

The differential absorption spectra obtained in the presence of these two nucleotides are indeed similar to those obtained after reduction electrolysis of the complex in the first reduction wave, and obtained by pulse radiolysis. The presence of the deprotonated radical cation $GMP(-H)^{\bullet}$ can also be detected by recording the transient absorption after reaction of the reduced complex with O_2.

$$[Ru(TAP)_2(TAP^{\bullet -})]^+ + O_2 \longrightarrow Ru(TAP)_3^{2+} + O_2^{\bullet -} \quad (15)$$

In the absence of oxidants (O_2 or other), the back electron transfer between the reduced complex and the radical of the base takes place.

The formation of the transient reduced complex can also be observed by flash photolysis in the presence of CT-DNA [100]. This shows clearly the existence of a photoinduced electron transfer from a base of the polynucleotide to the excited complex. However, the relative amount of reduced complex which is photoproduced, is smaller in the presence of CT-DNA than in the presence of GMP; this may be attributed to a more important back electron transfer process in the ion pair produced on the polynucleotide compared to that in solution with the mononucleotide.

Interestingly, the correlation of the luminescence quenching by the mononucleotides and polynucleotides with the occurrence of a photoelectron transfer process examined by flash photolysis, has been found [73, 95] for the whole series of Ru(II) complexes with TAP and HAT ligands discussed in Sect. 4.1. As will be developed in Sect. 5., this photoelectron transfer with polynucleotides is also connected to enhanced yield of strand breaks and the appearence of adducts on DNA.

It should be stressed that for the TAP and HAT Ru(II) complexes, although their modes of binding has been examined by luminescence spectroscopy, the nature of the excited species responsible for the photo-electron transfer (i.e. intercalated or surface bound) and the site specificity of these photoreactions, is not known at present.

4.3.2 Photo-Electron Transfer from or to an Excited Complex Interacting with DNA, and a Quencher

The influence of DNA on the photo-electron transfer process between a variety of donor-acceptor couples has been examined during the last ten years. For all the systems studied, the metal complex interacts with the DNA and plays the role of electron acceptor or donor in the hydrophobic DNA microenvironment, whereas the other partner of the process, i.e. the reducing or oxidising agent in the ground state, is localised either on the DNA double helix, or does not interact with the nucleic acid and remains in the aqueous phase. Thus three types of quenchers have been used for studying the influence of DNA on the rate of the electron transfer process:

– positively charged oxidative quenchers such as $Co(phen)_3^{3+}$ or $Cr(phen)_3^{3+}$ which bind physically to DNA;

– the negatively charged $Fe(CN)_6^{4-}$ reductive quencher which remains in the aqueous phase because of the electrostatic repulsion with the phosphate backbone;

– and neutral quenchers such as hydroquinone, quinone, or oxygen, which do not bind to DNA and are also located in the bulk aqueous solution.

The aims of such photo-electron transfer studies are manifold: (1) as mentioned in Sect. 3.3, the electron transfer process can be used to characterise the modes of binding of the Ru(II) complexes to DNA (differential quenching), (2) via this primary process, intermediates that react with the DNA bases or sugar and induce cleavage can be generated, (3) one can take advantage of the DNA double helical structure as molecular scaffolding for mediating the electron transfer between a donor and acceptor physically bound to DNA. We will consider successively these three applications.

4.3.2.1 Characterisation of Binding Modes of Ru(II) Complexes to DNA

Much information can be obtained on the interaction modes of Ru(II) complexes with DNA by studying the luminescence quenching by electron transfer with a negatively charged species which does not interact with DNA. Thus Barton and co-workers [51, 52] have examined from Stern-Volmer plots in intensity and lifetimes, the quenching of excited $Ru(phen)_3^{2+}$ by $Fe(CN)_6^{4-}$. As two luminescence lifetimes are observed in the presence of DNA without quencher, two Stern Volmer plots, yielding two different quenching rate constants k_q, are determined. The comparison of these two k_q values to that without DNA shows that one complex species is quasi-unquenchable by $Fe(CN)_6^{4-}$ (i.e. well protected from the surrounding aqueous solution) whereas the other Ru species has a quenching rate constant slightly lower than that in solution (i.e. less well protected from the external aqueous phase). These results have led the authors to infer the presence of a "partially intercalated" and "surface bound" Ru(II) species (Cf Sect. 3.4).

The protection of a reactive intermediate complex by the DNA double helix versus a neutral oxidising agent in solution, has also been demonstrated by studying a photo-electron transfer process. In this example the intermediate complex is produced photochemically on the DNA, and is examined spectroscopically after a laser pulsed excitation [73]. Thus $Ru(TAP)_2(HAT)^{2+}$ physically bound to nucleic acid is photo-reduced by hydroquinone during the laser pulse. The intermediate $[Ru(TAP)_2(HAT)]^+$ so-produced, detected by its absorption at 480 nm, is reoxidised by benzoquinone purposely added as oxidant to the solution. It is shown that this reoxidation of the mono-reduced complex is slower in the presence of polynucleotide than in its absence, indicating a protection of the transient mono-reduced complex in the DNA grooves.

Another interesting approach where electron transfer processes have been applied to examine the interaction and to determine the affinity constant of the complex for DNA is the electrochemical method [82] where the electrochemiluminescence (ECL) of $Ru(phen)_3^{2+}$ has been measured. The system is composed of $Ru(phen)_3^{2+}$ plus oxalic acid, in the presence or absence of DNA. It is well known that the electrochemical oxidation of $Ru(phen)_3^{2+}$ in the presence of oxalic acid produces the very reducing intermediate radical anion, $CO_2^{\cdot -}$ by the reaction:

$$Ru(phen)_3^{3+} + C_2O_4^{2-} \longrightarrow Ru(phen)_3^{2+} + CO_2 + CO_2^{\cdot -} \tag{16}$$

In the absence of DNA, the ECL of the complex is generated by reduction of the oxidised $Ru(phen)_3^{3+}$ by $CO_2^{\cdot -}$:

$$Ru(phen)_3^{3+} + CO_2^{\cdot -} \longrightarrow Ru(phen)_3^{2+*} + CO_2 \tag{17}$$

$CO_2^{\cdot -}$ is also capable of reducing the starting $Ru(phen)_3^{2+}$ complex:

$$Ru(phen)_3^{2+} + CO_2^{\cdot -} \longrightarrow Ru(phen)_3^{+} + CO_2 \tag{18}$$

The reaction of the reduced complex with the oxidised one may also produce excited $Ru(phen)_3^{2+}$:

$$Ru(phen)_3^{+} + Ru(phen)_3^{3+} \longrightarrow Ru(phen)_3^{2+*} + Ru(phen)_3^{2+} \tag{19}$$

In the presence of DNA reactions (17) and (18) that generate the excited complex directly or indirectly via reaction (19), become much slower or do not take place, and therefore the ECL disappears. This is due to the fact that the Ru(II) and Ru(III) complexes, physically bound to DNA, are protected by the negatively charged phosphate backbone from the reduction by $CO_2^{\cdot -}$. Thus the ECL titration of the metal complex in the presence of DNA has allowed the determination of the equilibrium constant and binding-site size for association of $Ru(phen)_3^{2+}$ to DNA [82].

4.3.2.2 *Production of DNA-Reactive Intermediates*

The best example illustrating this aim, is the case of the photosensitised reduction of persulfate by $Ru(bpy)_3^{2+}$ in the presence of DNA [105–108]. It is well known that the excited $Ru(bpy)_3^{2+}$ is able to transfer an electron to persulfate, with generation of the strongly oxidising intermediate species, the anion radical $SO_4^{\cdot -}$ [109]

$$Ru(bpy)_3^{2+*} + S_2O_8^{2-} \longrightarrow Ru(bpy)_3^{3+} + SO_4^{\cdot -} + SO_4^{2-} \tag{20}$$

As the quencher is negatively charged, this electron transfer reaction and the subsequent reactions involving the negatively charged $SO_4^{\cdot -}$ radical, are less efficient when the complex binds to DNA than when it remains in solution, nevertheless $SO_4^{\cdot -}$ and the oxidised complex oxidise the bases (B) of nucleic acid, eventually leading to strand scissions (see Sect. 5).

$$Ru(bpy)_3^{2+} + SO_4^{\cdot -} \longrightarrow Ru(bpy)_3^{3+} + SO_4^{2-} \tag{21}$$

$$Ru(bpy)_3^{3+} + B \longrightarrow Ru(bpy)_3^{2+} + B^{\cdot +} \tag{22}$$

$$SO_4^{\cdot -} + B \longrightarrow SO_4^{2-} + B^{\cdot +} \tag{23}$$

Another example which can also be reported in this section is the case of $Ru(phen)_3^{2+}$ and derivatives, which photosensitize DNA cleavages in the presence of oxygen [49]. Although the primary process is generally regarded as an energy transfer from the excited complex to oxygen [49], which produces singlet oxygen attacking the bases, the possibility of a quenching by electron transfer could also be considered [110]:

$$Ru(phen)_3^{2+*} + O_2 \longrightarrow Ru(phen)_3^{3+} + O_2^{\cdot -} \quad (24)$$

The Ru^{3+} complex which is formed is able to react with DNA components such as guanine bases [105–108]. It has indeed been demonstrated that DNA fragments and hepta decanucleotides with a hairpin structure treated with $Ru(bpy)_3^{3+}$ and subsequently with piperidine, give rise to cleavages at guanine residues in the single-stranded region [111]. It is also possible that $O_2^{\cdot -}$ could be converted to reactive $OH^{\cdot}$ species via the formation of H_2O_2 [112].

4.3.2.3 DNA as Molecular Scaffolding for Photo-Electron Transfer Studies
Several studies have been devoted to the dynamics of photo-electron transfer processes between a donor and an acceptor both interacting with the DNA double helix. These studies aimed at analysing the influence of the double helical structure of DNA on the electron transfer rate. For example, one might wonder whether the phosphate backbone or the stacking of bases could, in some fashion, mediate long range electron transfer and whether the dynamics of this process would be different for intercalated or surface bound complexes.

To clarify these problems, Stern Volmer constants (K_{SV}) have been determined from plots of luminescence intensity for the quenching of $Ru(phen)_3^{2+}$ by Co(III) complexes of "phen", "bpy", and "DIP" [113]. From these first experiments, it was found that the K_{SV} values are higher when the quenching takes place on the DNA. In order to be able to differentiate some of the possible couples of redox reagents (for example, an intercalated or surface bound excited Ru(II) complex and the quencher), the Stern Volmer constants and quenching rate constants (k_q) have been determined from emission lifetimes measured by single photon counting [114]. From these experiments it is possible to determine the lifetimes of the surface bound and intercalated excited Ru(II) complex and hence the corresponding quenching rate constants, k_{qs} and k_{qi} respectively. The k_q value associated with the quenching of the surface bound Ru(II) complex is somewhat higher than the one associated with the quenching of the intercalated Ru(II) compound. Both values are greater than the corresponding k_q constant in the absence of DNA, and it was therefore concluded that DNA accelerates electron transfer.

From experiments at low temperature where the donor and acceptor would be "frozen" on the DNA and thus not able to diffuse along the nucleic acid, it was suggested that a long range, DNA-mediated, photo-electron transfer, was taking place [114] between $Ru(phen)_3^{2+*}$ and different oxidising metallic complexes. This hypothesis was suggested although the emission decay curves which are shown, are biexponential for the whole concentration range of quencher in glycerol/water, conditions where stretched exponentials would have been expected.

It should be mentioned that, although the k_q values are always higher in the presence of DNA at room temperature, it is possible that this can be explained in ways other than an accelerating effect of the DNA on the electron transfer process. This has been shown for example with the system $Ru(TAP)_2(HAT)^{2+}$/ethidium bromide with poly[d(A-T)]. In this case, the electron transfer takes place from the ethidium to the excited complex [81] (energy transfer processes in this system have been shown to be absent on DNA). This donor-acceptor couple presents some advantages which allow some simplifications as compared to the case of $Ru(phen)_3^{2+}$. The ethidium bromide (the quencher) is known to be intercalated under the experimental conditions. Moreover, as the excited state lifetime of the "partially intercalated" Ru(II) species can be determined from single photon counting experiments, one has a direct access to the k_q value for the quenching of the excited "partially intercalated" complex by the ground state of the intercalated ethidium. For the electron transfer between these two intercalated species, it has been shown that the apparent quenching rate constant is again higher in the presence of polynucleotide (poly[d(A-T)]). After a correction for the DNA effect on the concentration of the reactants on the nucleic acid (based on a model proposed by Fromherz and Rieger. [115]), it is concluded that the quenching rate constant k_q is in fact 10^3 times lower on the polynucleotide than in solution. This may be attributed to a decrease of mobility of the complex along the DNA, due to its "partial intercalation" into the double helix. Alternatively, we suggest that, as it is not certain whether a "partially" intercalated complex can really diffuse along the DNA during its excited state lifetime, the excited "partially intercalated" species might be immobilised into the stacking of bases during its lifetime; consequently this lower rate constant for the electron transfer could be attributed to a slow transfer of the electron between immobile donor and acceptor on the nucleic acid. Such electron transfers were reported recently by Brun and Harriman for organic intercalated donors and acceptors [116].

5 Photoreactions of DNA Initiated by Ru Polypyridyl Complexes

Upon photo-excitation Ru(II) complexes have been reported to induce three main kinds of reactions of DNA: strand cleavage, nucleobase modification, which leads to strand breaks on subsequent treatment with base, and photo-adduct formation.

5.1 Strand Cleavage

The induction of single strand breaks in DNA can be conveniently studied using plasmid DNA. "Nicking" of a bond in one strand of the duplex of the supercoiled closed circular form of this DNA causes the molecule to convert to its open circular form, a change that is very readily monitored by gel electrophoresis. When Ru(II) complexes are excited by visible light in the presence of samples of supercoiled plasmid DNA, single strand breaks are observed. The reaction yield is diminished by increasing the ionic strength of the solution, indicating that binding to the DNA is necessary for efficient reaction. As both $Ru(phen)_3^{2+}$ and $Ru(bpy)_3^{2+}$ are active, it appears that the mode of binding is not critical for the process to take place [49, 74, 117]. Although the cleavage reaction is readily monitored in a short time because of the sensitivity of the plasmid assay method, the quantum yield is in fact rather low. Thus for $Ru(bpy)_3^{2+}$ in air-saturated aqueous solution a quantum yield of 6.6×10^{-6} has been determined [105]. This is reduced to 1.2×10^{-6} if the solution is argon-degassed, indicating that there are both oxygen-dependent and oxygen-independent reaction pathways.

A role for singlet oxygen is implied by an enhancement of cleavage in D_2O and by the suppression of cleavage by $^1O_2$ quenchers, histidine and azide ions [49, 117]. However whether singlet oxygen is capable of inducing frank single strand breaks is still controversial [118]. The cleavage, which occurs at guanine [119], could possibly arise at particularly sensitive oxidised nucleobase sites. Alternatively it may be that the reaction is due to different reactive oxygen species, with hydroxyl radical type reactivity. These might be formed from singlet oxygen or from the superoxide (accompanied by formation of Ru(III) complexes) generated by photoinduced electron transfer (Eq. 24, Sect. 4.3.2).

Singlet oxygen is well-known to cause photo-oxidation of guanine bases in DNA [118]. The site of this damage is revealed after a subsequent treatment of the polynucleotide by a base (e.g. piperidine at 90 °C), which induces breaks in the DNA. Such alkali-labile sites are usually associated with nucleobase modification or removal. Singlet oxygen is known to add to guanine forming initially endoperoxides and subsequently species such as 8-hydroxyguanine [118]. For several Ru polypyridyl complexes, formation of such alkali-labile sites has been demonstrated using ^{32}P-end-labelled oligonucleotides or restriction fragments. It is found that the damage occurs selectively at guanine bases [119, 66], with some preference for sequences CGA, TGA and CGT [119]. No quantum yields for this process have been reported but the reaction is certainly much more efficient than the formation of direct strand breaks. Attachment of a $Ru(phen)_3^{2+}$-group to a short oligonucleotide allows the targeting of the photochemical reaction to neighbouring sites on a complementary single strand [119].

Mei and Barton [66, 67] have used photo-sensitised damage to DNA by Λ-$Ru(TMP)_3^{2+}$ (induced by $^1O_2$ formation) as a means of probing A-DNA conformations, and they have compared the damage so caused on a linear pBR322 to that produced by $Ru(phen)_3^{2+}$, which will preferentially bind to B-DNA sections. For example it was found using tritiated polynucleotides that irradiation of Λ-$Ru(TMP)_3^{2+}$ leads to cleavage of poly(rC).poly([^{3}H]dG) (A-form double helix) but not poly([^{3}H]dG-dC) (B-form). Using ^{32}P-labelled restriction fragments of plasmid DNA, it was shown that Λ-$Ru(TMP)_3^{2+}$, as well as causing cleavage at guanine, also reacts with homopyrimidine stretches. This is attributed to preferential attack at the sites where Λ-$Ru(TMP)_3^{2+}$ is bound.

5.2 Photocleavage Following Photo-Induced Electron Transfer

While most ruthenium complexes studied show only modest reactivity in cleaving DNA [74] significantly higher activity is found for $Ru(TAP)_3^{2+}$ [74, 96, 97]. Moreover there exists a correlation between the photocleaving ability by a series of TAP and HAT complexes and their oxidation power in the excited state. As indicated above in Sect 4.3., the excited state of these complexes undergoes electron transfer and it is reasonable to assume that the enhanced yield of cleavage is due to reactions of the guanine radical cation. Nucleobase radical cations [120] (or their deprotonated derivatives [121]) are known to abstract H-atoms from nearby riboses [122]. The radical of the sugar will then induce the break of a phosphodiester bound, possibly similar to that found with ionizing radiation [123].

Efficient direct cleavage of DNA has been observed when $Ru(bpy)_3^{2+}$ is photo-excited in the presence of persulfate ion [105–109]. The quantum yields are three orders of magnitude higher (e.g. 0.0084 at a phosphate/complex of 18) than for $Ru(bpy)_3^{2+}$ alone. The yields of the reaction are reduced in the presence of air, by addition of mono- or divalent ions, and when $Ru(phen)_3^{2+}$ replaces $Ru(bpy)_3^{2+}$. As discussed above (Sect. 4.3.2) the initial photochemical reaction is a photo-induced electron transfer yielding two strongly oxidising species, $Ru(bpy)_3^{3+}$ and $SO_4^{\cdot -}$ [105–108]. Reaction of bases with the Ru(III) species is however much slower than with $SO_4^{\cdot -}$. The biological activity of DNA photolysed in the presence of $Ru(bpy)_3^{2+}/S_2O_8^{2-}$ was determined by examining the survival of *E. coli* bacteria after transformation of the plasmid. These experiments indicated that the quantum yield for deactivation was high (about 80% of that of strand breaks), although it appears that more than 30 $SO_4^{\cdot -}$ radicals per plasmid are required to induce one lethal event [105].

5.3 Photoadduct Formation

Gel electrophoresis of ^{32}P-end-labelled oligonucleotides irradiated with visible light in the presence of $Ru(TAP)_3^{2+}$ showed that the principal photochemical product is a less electrophoretically-mobile species [96]. This is consistent with the formation of a photo-adduct and it is clear that the yield of this reaction is much greater than that for the induction of direct strand breaks (see above) or for the formation of alkali-labile sites. Adduct formation is also observed with $Ru(TAP)_2(bpy)^{2+}$ but not with $Ru(TAP)(bpy)_2^{2+}$ [96]. Further evidence for adduct formation both for single and double-stranded DNA has been obtained by dialysis and UV/visible spectroscopy [124]. Photoadduct formation with DNA is insensitive to solution pH or aeration. Experiments with poly[d(G-C)] and poly[d(A-T)] indicate that the adduct is probably formed with guanine, and this is supported by the similar spectroscopic changes observed with GMP under certain conditions [124]. This, as well as other observations, indicates that the adduct would result from an electron transfer from a base (probably a G) to the excited complex. It is postulated that the adduct is formed after this photo-electron transfer, by combination of the $[Ru((TAP)_2(TAP^{\cdot -})]^+$ and oxidised guanine, possibly after proton transfer:

$$Ru(TAP)_3^{2+}, G \xrightarrow{h\nu} [Ru(TAP)_2(TAP^{\cdot -})]^+, G^{\cdot +} \tag{25}$$

$$[Ru(TAP)_2(TAP^{\cdot -})]^+, G^{\cdot +} \longrightarrow [Ru(TAP)_2(TAPH^{\cdot})]^{2+}, G(-H^+)^{\cdot} \tag{26}$$

$$Ru(TAP)_2(TAPH)^{2+}, G(-H^+)^{\cdot} \longrightarrow \text{Photo-adduct.} \tag{27}$$

Further experiments will need to be carried out in order to determine the factors that induce cleavages versus adducts, as both reactions are apparently initiated by a photo-electron transfer process.

6 Polypyridyl Rh(III) and Co(III) Complexes with DNA

6.1 Photophysics of Rh(III) Polypyridyl Complexes

6.1.1 Tris-Polypyridyl Complexes

The absorption spectra of tris-polypyridyl Rhodium(III) complexes are characterised by several intense Ligand Centered (LC) absorption bands in the UV. Neither MC absorption bands, nor CT bands are observed in the visible region of the spectrum in contrast to their Ruthenium analogues. This makes tris(polypyridyl)Rh(III) complexes formed with bpy and phen practically colorless [1].

At low temperature, the emission spectra of the complexes are well-structured and assigned to "ligand-localised $\pi\pi^*$ phosphorescences", responsible for multiexponential luminescence decays observed with mixed-ligand compounds [125, 126].

Emissions from both a MC and LC excited state were observed at low temperature with sterically hindered ligands such as 3,3′-Me_2-bpy [127] and 2,2′:6′,2″-terpyridine [128]. The MC emission is the dominant feature at 77 K, but the LC emission is enhanced relative to the metal centred one in fluid solution [127].

The behaviour in room-temperature fluid solutions of excited Rhodium(III)-polypyridyl complexes remains unclear. These compounds are weak emitters, and perhaps because of this, contradictory reports on the room temperature emissions of $Rh(bpy)_3^{3+}$ and $Rh(phen)_3^{3+}$ have been published. Indelli et al. [129] detected the emission at 588 nm (dd*) and 455 nm ($\pi\pi^*$) for $Rh(phen)_3^{3+}$ while Nishizawa et al. [127] observed only the $\pi\pi^*$ emission at 455 nm. The tris-polypyridyl Rhodium(III) complexes photodissociate, giving rise to the loss of a ligand [130], as is expected when the MC state can be populated.

$$[Rh^{III}(LL)_3]^{3+} + 2H_2O \xrightarrow{h\nu} [Rh^{III}(LL)_2(H_2O)_2]^{3+} + LL \qquad (28)$$

6.1.2 Bis-Polypyridyl Complexes

The bis-chelated complexes such as $Rh(phen)_2Cl_2^+$ or $Rh(bpy)_2Cl_2^+$ show an absorption, intense intraligand transitions in the UV region and a much less intense longer wavelength shoulder, which suggests an assignment to a MC transition [1]. They emit moderately at low temperature ($\lambda^{em} \cdot Rh(bpy)_2Cl_2^+ \approx 704$ nm [127]) from a dd (or MC) triplet state [131].

As expected for an MC state, the photoaquation is an efficient process [132]

$$[Rh^{III}(LL)_2X_2]^+ + H_2O \xrightarrow{h\nu} [Rh^{III}(LL)_2(X)(H_2O)]^{2+} + X^-$$

$$(LL = \text{bpy, phen, } X = \text{Cl, Br, I}) \qquad (29)$$

6.2 Photoredox Reactions of Rh(III) Polypyridyl Complexes

On the basis of the reduction potential of $Rh(phen)_3^{3+}$ ($E_0 = -0.75$ V/SCE) and of its $^3\pi\pi^*$ energy (2.75 eV), $Rh(phen)_3^{3+}$ in the $^3\pi\pi^*$ state is expected to be a very powerful oxidising agent (with a reduction potential of ≈ 2.0 V/SCE [133]), making it a stronger oxidant than the ^{3}MLCT states of the Ru(II) complexes discussed above. Electron transfer from aromatic amines [134] or di- and tri-methoxybenzenes [135] to excited Rh(III) polypyridyl complexes have indeed been observed.

Although the redox potentials of $Rh(phen)_2Cl_2^+$ are unknown, electron transfer from methoxybenzene to the ^{3}MC state of this complex has also been reported [134] however, with a lower quenching rate constant than for $Rh(phen)_3^{3+}$.

6.3 Photoreaction of Rh(III) Complexes with DNA

Bis- and tris polypyridyl Rh(III) complexes which have been studied with DNA can be found in Table 2.

6.3.1 DNA Cleavage

It has been shown that polypyridyl Rh(III) complexes induce photo-cleavages of the sugar phosphate backbone of double-stranded DNA with a higher relative quantum yield than Ru(II) complexes of phen or DIP. Thus replacement of Ru(II) ions by Rh(III) in Tris(phen) complexes, increases the efficiency of DNA photo-cleavages. However, in contrast to the Ru(II) complexes, Rh(III) samples have to be illuminated in the UV because of the absence of absorption bands in the visible region.

In order to determine the photocleavage mechanism by Rh(III) complexes, Barton [136] examined the photo-cleavage pattern of DNA, produced by complexes based on the Rh(phi) chromophore (phi: 9,10-phenanthrenequinone diimine, Fig. 2), $Rh(phi)_2(bpy)^{3+}$ and $Rh(phi)(phen)_2^{3+}$ (Fig. 2). The cleavage appears to involve a photo-generated species which induces strand scission of the DNA backbone at its binding site. The analysis of the DNA degradation products resulting from the photoreactions of $Rh(phen)_2phi^{3+}$ is consistent with the initial step being an abstraction of hydrogen from the C'_3 position of the sugar. The resulting radical would lead to strand scission and base release.

It would be interesting to test with other Rh(III) complexes, whether the direct oxidation of the base (by photo-electron transfer) could also be a primary step responsible for photocleavages. Indeed, as outlined before in Sect. 5, radiation studies have shown that the radical cation of the base can produce the sugar radical, itself leading to strand scission [122]. Moreover base release, as observed with the Rh(III) complexes, can also take place from the radical cation of the base [137]. Direct base oxidation and hydrogen abstraction from the sugar could be two competitive pathways leading to strand scission and/or base release.

The recognition properties of the tris-chelate complexes with phen, phi, or DIP ligands (see Sect. 3.4) for different DNA structures and conformations, combined with the photoreaction of the corresponding Rh(III) complexes, have been used for the design of shape selective photocleaving agents. Thus $Rh(DIP)_3^{3+}$ has been used as a specific probe of DNA cruciform [138] where it photosensitises the cleavage of both DNA strands. $Rh(phen)_2phi^{3+}$ recognises changes in base-pair propellor twisting of double-helical DNA [139, 140, 136] due to steric interactions of its ancillary phenanthroline ligands. Moreover this same complex also photo-cleaves t-RNA at bases that are involved in triple interactions, in which normal Watson-Crick base pairs interact with a third base in the major groove [141].

6.3.2 Adduct Formation

The loss of ligand upon illumination for Rh(III) complexes might allow their anchoring to DNA. Morrison and co-workers [142, 143] have reported the photoreaction of $Rh(phen)_2Cl_2^+$ in the presence of mononucleotides or DNA. Under illumination, the loss of a chloride ligand generates a reactive species which binds to a DNA base, leading to the formation of adducts with a quantum yield of approximatively 10^{-3}. Thus covalent binding of the complex to guanine (probably via the N1 and N3) has been observed and an adenine adduct has also been isolated. When tris-chelate Rh(III) complexes are illuminated in the presence of DNA adduct formation has also been reported [136]. In this case, although no mechanistic studies have been performed, two possible pathways may be considered: (1) the dechelation of the complex via population of the MC state, with production of a reactive species able to bind to a nucleic acid base; (2) the photo-oxidation of a base, via a photo-electron transfer, with formation of a radical pair which leads, as proposed with $Ru(TAP)_3^{2+}$, to the adduct.

6.4 Photochemistry of Co(III) Complexes

The photochemical behaviour of Co(III) complexes is characterised by the appearance of photoaquation and photoredox reactions, depending on the nature of the lowest electronic transition [144, 145]. The intense band which appears in the ultraviolet spectral region is due to spin-allowed LMCT transition (Ligand to Metal Charge Transfer), for example in $Co(NH_3)_6^{3+}$. In mixed-ligand complexes, the low energy LMCT bands bands are those which involve the most reducing ligand(s). Therefore for those complexes, irradiation in the LMCT bands tends to induce redox decomposition, yielding Co^{2+} and oxidised ligand. On the other hand, irradiation in the *d-d* absorption bands (or MC) induces photosubstitution (such as photoaquation).

In most complexes, the charge transfer band is well separated from the MC band, so that it is easy to obtain excited states of different nature by exciting with radiation of suitable wavelengths. $Co(phen)_3^{3+}$ and $Co(bpy)_3^{3+}$ [146] do not present charge transfer absorption bands (LMCT or MLCT) in the visible but only a weak MC band and π-π^* transitions in the UV.

$Co(phen)_3^{3+}$ and $Co(DIP)_3^{3+}$ have been reported to cleave DNA upon irradiation with UV light ($\lambda < 320$ nm) [117, 147]. As no mechanistic studies were performed, the different reactions leading to strand scissions are not known. Photoreduction of the Co(III) species could constitute the initial step of the reaction pathway.

The same stereospecific interaction found with $Ru(phen)_3^{2+}$ or $Ru(DIP)_3^{2+}$ enantiomers with DNA of different helicities, has been observed in the photocleavage reactions by the corresponding Co(III) enantiomers, as indicated by the specific cleavage of left-handed DNA by Λ-$Co(DIP)_3^{3+}$ [148]. The use of

Λ-$Co(DIP)_3^{3+}$ as a probe of Z-DNA conformation in DNA genome [149] and in in vivo DNA [150] were also reported.

$Co(NH_3)_6^{3+}$ also photo-cleaves DNA [117], but in this case, formation of Co(II) and oxidised ligand from the LMCT state could represent an alternative pathway to a direct oxidation, leading to strand scission.

It has also been reported that Co(III)-bleomycin (or synthetic analogues) cleave DNA when illuminated with UV [151–153].

7 Photophysics and Photochemistry of Cationic Porphyrins and DNA

The study of porphyrin-sensitised reactions with biological systems has been greatly stimulated by the applications of porphyrins for photodynamic therapy of tumours. In this treatment, for which initially haematoporphyrin derivatives and purified fractions (such as Photofrin II) have been used, DNA does not appear to be the principal target. There is, nevertheless, clear evidence that these porphyrins can induce DNA-damage upon photo-irradiation, which could lead to genetic toxicity [154]. In this section, however, we will concern ourselves predominantly with cationic porphyrins and in particular with *meso*-tetrakis (4-*N*-methylpyridiniumyl)porphyrin [155] derivatives (Fig. 13, **1**) where detailed binding studies and photophysical measurements have been reported (for reviews see [155–157]). Following on from the pioneering work of Fiel it has been shown by several groups that, despite its size, the free base H_2TMPyP^{4+} can intercalate into DNA. More recent detailed studies with a range of *meso*-substituted cationic porphyrins suggest that only half of the porphyrin ring is necessary for intercalation to occur [158]. Similar behaviour is found for the Pd(II) and Au(III) derivatives where the metal is in square planar coordination. With other metalloderivatives such as $ZnTMPyP^{4+}$ or $MnTMPyP^{5+}$, where ligands are present either in one or both of the axial coordination positions, intercalation is not possible, and the complex appears to be groove-bound.

The number of DNA-bound porphyrins, whose excited state properties have been studied, is quite small, partly because many metalloporphyrins emit only weakly, if at all [1, 159]. The fluorescence of H_2TMPyP^{4+} itself is strongly affected upon binding to DNA, the broad spectrum of the porphyrin being split into two bands [160]. Studies with synthetic polynucleotides show contrasting behaviour for poly[d(G-C)] and poly[d(A-T)]. With the G-C polymer the fluorescence quantum yield is reduced to ca.50% of that of the free porphyrin, whereas with the A-T polymer the quantum yield doubles [161]. The quenching with the guanine-containing DNA has been attributed to an electron transfer process as the excited state is estimated to be sufficiently oxidising [$E^0(P^*/P^-) = 1.60$ V/NHE] to form the guanine radical cation but not that of

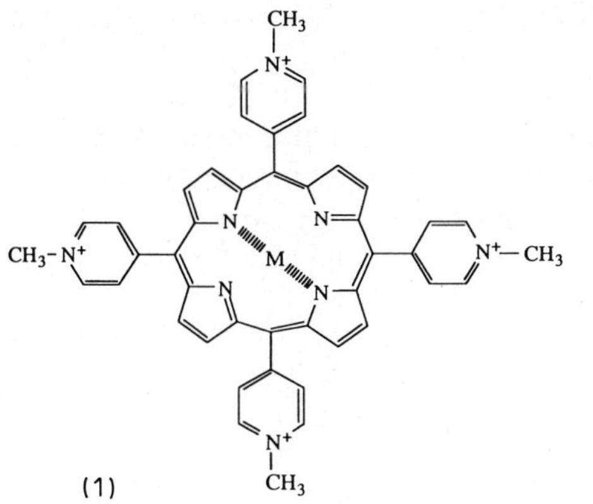

(1)

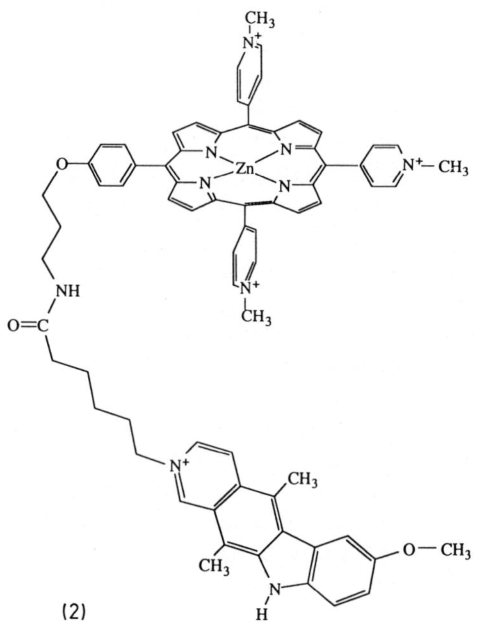

(2)

Fig. 13. Structures of $MTMPyP^{4+}$ (**1**) and $ZnTMPyP^{4+}$-R-ellipticene (**2**)

adenine or of the pyrimidines [160] (see Sect. 4.3.1).

$$P + G \xrightarrow{h\nu} P^{\bullet -} + G^{\bullet +} \tag{30}$$

Time-resolved measurements revealed two emitting sites for DNA-bound H_2TMPyP^{4+} with lifetimes of 1.7 and 10 ns [162]. The signals were ascribed to intercalated and externally-bound species. With $ZnTMPyP^{4+}$ the fluorescence spectrum is markedly affected upon binding to poly[d(A-T)] but not to poly[d(G-C)], probably indicating weak interaction with the G-C polymer and consistent with strong groove binding with poly[d(A-T)]. The lifetime of the singlet state of $ZnTMPyP^{4+}$ increases from 1.3 to 1.8 ns upon binding to DNA [162]. The triplet states of the DNA-bound porphyrins can also be readily monitored by laser flash photolysis [161, 163]. The triplet lifetimes of H_2TMPyP^{4+} and $ZnTMPyP^{4+}$ are lengthened on binding to DNA and the rate constant for deactivation by O_2 is markedly reduced, indicating the extent to which the polynucleotide shields the porphyrin.

UV excitation of the DNA bases, which leads to enhanced fluorescence for several cationic porphyrins including H_2TMPyP^{4+} through contact energy transfer, provides further evidence for intercalation of these species [164]. When both are bound to DNA, H_2TMPyP^{4+} is capable of quenching ethidium fluorescence at distances of 25–30 Å [165]. It is proposed that the excited state quenching proceeds by Förster-type energy transfer.

Compound (**2**) (Fig. 13) containing covalently-linked $ZnTMPyP^{4+}$ and ellipticene binds very strongly to DNA ($K > 10^8\ M^{-1}$), the ellipticene part intercalating while the porphyrin moiety is groove-bound [166]. Binding to DNA greatly enhances the fluorescence yield and singlet oxygen yield for (**2**) presumably by separating the porphyrin and ellipticene moieties and consequently preventing the deactivation of the excited singlet and triplet states.

Excited state resonance Raman spectra of $CuTMPyP^{4+}$ bound to DNA or poly[d(A-T)] have been recorded [167, 168]. These are assigned to an exciplex formed between the porphyrin and the A-T sites of the polynucleotide. The excited state lifetime is estimated to be ca. 20 ps. Weak emission from $CuTMPyP^{4+}$ bound to DNA has been reported and has been assigned to originate in a tripdoublet or tripquartet level [169]. It is believed that the emissive complexes are intercalated, whereas groove-bound $CuTMPyP^{4+}$ does not emit because of solvent quenching of the excited state.

Although certain metalloporphyrins, notably iron-derivatives (in the presence of oxygen and a reducing agent) [170] or manganese derivatives (in the presence of oxidising agents such as persulfate) [171] are effective cleavers of DNA [172], such compounds are not efficient photosensitisers of the cleavage of DNA. This is presumably a consequence of the short lifetimes of their excited states. Photochemical cleavage of DNA has been achieved using H_2TMPyP^{4+}, $ZnTMPyP^{4+}$ and $PdTMPyP^{4+}$ [160, 163, 173a, 174]. Such cationic porphyrins are much more efficient at cleaving DNA than are anionic porphyrins, presumably indicating that binding to the polynucleotide is necessary for high yield. A more recent study [175] with a range of substituted cationic porphyrins indicates that those dyes which intercalate are more effective at cleaving plasmid DNA and that the photosensitiser efficiency increases with the number of positive charges on the porphyrin. Surprisingly therefore Munson and Fiel [173b] have reported that both *cis-* and *trans-* bis(N-methyl-4-pyridiniumyl)diphenylporphyrin are more effective at inducing strand breaks than H_2TMPyP^{4+}. Praseuth et al. [174] showed that there is a rough correlation between the effectiveness of the cationic porphyrins in sensitising direct strand breaks in aerated solution and their ability to sensitise the formation of singlet oxygen. A role for this species is supported by the partial quenching of the reaction by azide ion, although some reaction appears also to proceed by an anaerobic pathway. The yield of strand breaks was not increased using a bi-porphyrin analogue of H_2TMPyP^{4+}. Porphyrins also photosensitise the formation of alkali-labile sites at guanines in DNA, presumably through photo-oxidation via singlet oxygen [175] and this reaction proceeds in significantly higher yield than direct strand breaks. Irreversible sequence specific damage to DNA has been achieved using free base [176] or Pd-porphyrin [177] attached to a short oligonucleotide. Both the induction of alkali-labile and DNA cross-linking were observed. Despite its high binding affinity the ellipticene-linked porphyrin (**2**) has been shown to be no more effective at inducing single strand breaks in plasmid DNA than $ZnTMPyP^{4+}$ (quantum yield for both is 2×10^{-6}), although it is more than 50 times more effective than HPD [166b].

Fluorescence studies indicate that both H_2TMPyP^{4+} and $ZnTMPyP^{4+}$ are able to penetrate through to the nuclei of plant cells and are taken up by the chromatin [178, 179]. Both porphyrins can sensitise photo-damage to the DNA, the genotoxic effect being greater for $ZnTMPyP^{4+}$. An increase in fluorescence from H_2TMPyP^{4+} is observed upon light exposure of rat epithelial cells containing the porphyrin [180]. The effect is particularly marked for fluorescence in the cell nucleus. Lambda phage, a non-enveloped double helical DNA protein virus, can be inactivated by photolysis with H_2TMPyP^{4+} both in the presence and absence of air [181]. It is proposed that the reaction occurs via guanine oxidation following photo-induced electron transfer (reaction 30). $PtTMPyP^{4+}$ also rapidly causes viral inactivation, although in this case the reaction was oxygen dependent.

8 Photochemistry of Uranyl ion and DNA

The photochemistry of uranyl ion has been very extensively studied [182, 183]. The compound has been found to be an extremely effective oxidiser and hydrogen abstractor, such behaviour being shown with a wide variety of organic compounds and ligands. The excited states are quite long-lived; a biexponential decay is observed in emission or transient absorption in aqueous acidic solutions (due to two excited states, average lifetime ca. 2.4 μs). It is also instructive to note that the excited state is a strongly oxidising species (Excited state energy in H_2O; 246 kJ mol^{-1}; $E(U^{VI}/U^{V}) = 0.05$ V/NHE; $E(U^{VI*}/U^{V} = 2.60$ V/NHE). It may cause the abstraction of hydrogen atoms from organic compounds such as alcohols, however, if this proceeds by direct H-atom abstraction or by initial formation of a hydroxyl radical from coordinated water [184] remains unclear.

$$U^{VI}O_2(H_2O)_n^{2+*} \longrightarrow [U^{V}O_2(H_2O)_{n-1} - H^{\bullet}, OH^{\bullet}]^{2+} \qquad (31)$$

Given that both hydrogen-abstraction and photo-oxidation would be expected to lead to DNA-damage, it seems most appropriate to use uranyl ion to induce photochemical damage on DNA. It may be noted however that uranyl ion is only stable at pH < 3; at higher pHs the complex hydrolyses and aggregates forming polyuranates and insoluble hydrated oxides. Despite these problems Nielsen [185, 186] has recently found that uranyl ion is an extremely effective photosensitiser for DNA-cleavage, apparently acting by hydrogen abstraction (direct or through OH-like species) and causing damage which for random sequences of double-stranded DNA is essentially independent of the type of nucleotide – i.e sequence neutral. This therefore makes this system a most useful one for studying the structure and conformation of DNA complexes with proteins and other large molecules.

Based on kinetic measurements it has been estimated that uranyl ions have an extremely high affinity for DNA (K = ca. 10^{10} M^{-1} at pH 4) [186]. At the pHs used for biological experiments (typically 6 or 7) the DNA-uranyl complex is metastable. However the high affinity constant and the long lifetime of the complex allow it to be used in such experiments. Flow linear dichroism experiments indicate that the uranyl is groove-bound, possibly bridging the phosphates across the minor groove. Binding to the DNA is necessary for efficient photocleavage, and strand breaking takes place next to the group where the uranyl ion is bound. Cleavage appears to be equally probable at the 3′- or 5′-positions. Strand breaking, which has a quantum yield greater than 10^{-4}, is accompanied by base release. The mechanism has not yet been fully elucidated. When uranyl ion is excited in the presence of mixtures of the four nucleo-bases, guanine is preferentially destroyed (consistent with its relative ease of oxidation). However in DNA no photo-oxidative damage of the bases is found. This is presumably because the strong binding to the phosphates prevents access of the uranyl ion to the bases in the DNA duplex. It therefore appears most probable that the DNA damage is caused by H-abstraction from the ribose. It is not yet clear which H-atom on the ribose is attacked by the uranyl ion and indeed it is possible that there is more than one site of attack.

The sequence neutrality for random stretches of double-stranded DNA makes uranyl ion a very useful reagent for examining contact regions in protein-DNA complexes. Such photo-footprinting studies have been carried out with the λ-repressor/OR1 [185], *E. Coli* RNA polymerase/*deo* P1promoter [187] and transcription factor IIIA-ICR [188].

Although the strand cleavage is normally a sequence neutral reaction, runs of A-T appear to show enhanced reactivity [189]. The preferential reaction in this area is believed to be due to stronger binding of the UO_2^{2+} to such extended AT tracts because of the narrower minor groove and higher negative electrostatic potential. Uranyl ion has also been used to photofootprint triple stranded DNA (one of the sequences used is shown below) [190]. It was found that binding of the third strand afforded little protection to the pyrimidine strand against uranyl-induced damage but did provide substantial protection to the purine strand, especially at its 5′-end. This can be interpreted in terms of the accessibility of the phosphate groups of the two strands and is consistent with the model where the third strand (the T_{15} oligonucleotide in this case) is located closer to the purine strand than to the pyrimidine strand in the major groove.

T T T T T T T T T T T T T T T

5′-A G C T T A T A T A T A T A T A A A A A A A A A A A A A A A T C G A T A G G A T C-3′

3′-T C G A A T A T A T A T A T A T T T T T T T T T T T T T T T A G C T A T C C T A G-5′

9 Conclusions and Perspectives

While the area of the photochemistry of metal complexes with DNA is still at an early stage of development, substantial progress has been made in the last ten years. It is apparent from the different chapters of this review that a knowledge of the photophysics of the metal complex's excited state can be used to predict the type of reaction caused to the DNA.

MLCT ⟶ Redox Reaction. Cleavage[A] or Adduct Formation [124]

MC ⟶ Photosubstitution Reaction by a base [142]

LC ⟶ Redox Reaction (?) Cleavage (?)

LMCT ⟶ Redox Reaction and Hydrogen Abstraction [136]. Cleavage

A: 49, 74, 96, 97, 100, 119

For example for the HAT and TAP complexes of ruthenium discussed in Sect. 4.1, the excited state reduction potential and the accessibility of the ^{3}MC excited state correlate well as determining factors in the DNA reactions observed. If oxidation of the nucleo-base is possible (e.g. for guanine in the presence of $Ru(TAP)_3^{2+*}$) then enhanced yield of cleavage and adduct formation are found. However, these oxidising excited states are also photo-labile, because of the thermal population of the ^{3}MC state from the ^{3}MLCT. Where electron transfer from the base is not possible (e.g. for adenine in the presence of $Ru(TAP)_3^{2+*}$), ligand loss can occur, probably with the subsequent coordination of the base to the metal centre [95]. Similarly the formation of coordination-type photoadducts found with Rh(III) bis-polypyridyls appear to originate from ^{3}MC excited states [142]. In the case of other Rh(III) tris-polypyridyls, where the lowest excited state is expected to be of ^{3}LC-type [1], photocleavage is observed. Although the ^{3}LC state should be sufficiently oxidising to induce formation of the radical cation and subsequent strand breaks, it has been suggested that the reaction proceeds via H-atom abstraction by an LMCT state [136]. The LMCT state of $Co(NH_3)_6^{3+}$ and the LC state of $Co(DIP)_3^{3+}$ should lead to redox reactions in accordance with the observation of photocleavage [117].

H-atom abstraction has been demonstrated to be the mechanism of action of excited uranyl ions, and in this case negligible base oxidation is found. Nucleobase (especially guanine) oxidation is the principal reaction caused by singlet oxygen and this reactive species can be generated by a number of the complexes (e.g. many Ru(II)polypyridyls and porphyrins). It is worth pointing out, however, that the yield of $^1O_2$ may be lower when the sensitiser is bound to DNA, and it is the authors' view that some of the reactions claimed to proceed via $^1O_2$ may be caused by direct reaction of the photo-oxidised sensitiser with the DNA.

In our introduction we mentioned that there were three general areas where eventually it might be hoped to exploit the photochemical properties of metal complexes. These were (1) the use of the luminescence properties of the complexes as probes for DNA structure, base sequence or conformation; (2) The employment of photochemical strand breaks as a tool in molecular biology and (3) The development of phototherapeutic procedures.

Metal complexes have already been shown to have great potential as molecular probes for DNA structure. As described in Sects. 4 and 5, such studies are particularly advanced for ruthenium polypyridyls, where the complexes have been used to recognise selectively A-DNA, Z-DNA etc. A successful application will, of course, require a thorough knowledge of how such metal complexes bind to DNA and how this is affected by the conditions in the surrounding media. As discussed in detail above for the case of $Ru(phen)_3^{2+}$, this is not necessarily straightforward. However one of the considerable advantages of this type of complex is that control of the structure, and hence of the interaction with DNA, can be achieved by variation of the type of ligand (c.f. the use of DPPZ to facilitate intercalation). This, coupled with derivatisation to give site-specific binders (e.g. using oligonucleotide-bound reagents) [7, 8, 9], holds great promise for new reagents.

Several classes of metal complexes have been shown to induce chemical damage in DNA, for example the polypyridyls of ruthenium, rhodium and cobalt, metalloporphyrins and the uranyl ions. In particular the ability of these molecules to sensitise strand breaks in the DNA has been noted. However it is worth stressing that in many cases the quantum yield for the reaction is very low (10^{-6} is not uncommon). Particularly efficient systems so far reported include $Ru(bpy)_3^{2+}/S_2O_8^{2-}$ and UO_2^{2+}, where the quantum yield may exceed 10^{-3}. Moreover in most cases more needs to be known about the mechanism of DNA strand cleavage (e.g. whether it takes place through nucleobase radical cations, direct H-atom abstraction from ribose etc.). Further it is likely in many cases that strand breaks are only a minor photochemical reaction and it will be essential to determine the precise nature and yields of base-oxidation products. Another challenge in this area will be to improve the site-selectivity of binding, so as to more precisely direct the reagent to a particular sequence. The development of new oligonucleotide-bound reagents and other multifunctional compounds such as $Ru(TAP)_2POQ^{2+}$ [211] (Table 2, Fig. 2), seems particularly promising in this regard. This could open the way to the development of reagents having endonuclease-type activity. Here again it is essential to know in precise detail the processes leading to stand breaks. Ideally light-induced processes leading to catalysed hydrolysis of phospho-diester bonds without concurrent damage to the bases are desirable, but to the best of our knowledge no photochemical reactions of this type are known.

Finally, it is worthwhile summarising the potential of photochemical reactions on DNA induced by metal complexes as the basis of phototherapeutic reagents. It has been amply demonstrated that through oxidative damage both the metal polypyridyls and porphyrins are highly effective at reducing the biological viability of DNA. An alternative approach is to look for photochemical systems that form covalent adducts with DNA. It would be expected by analogy with the effects of drugs such as cis-platin $Pt(NH_3)_2Cl_2$ [13, 40, 41] and $Ru(DMSO)_4Cl_2$ [212] that such adducts would prevent the replication of tumour cells. Potentially, photochemical procedures have advantages in that the "drug" is not activated until bound to the DNA and illuminated; it might therefore be less toxic than other compounds. The observation of photoadducts from the $Rh(phen)_2Cl_2^+$ [142] and $Ru(TAP)_3^{2+}$ [96, 124] could be a starting point for this type of research.

10 References

1. Kalyanasundaram K (1992) Photochemistry of Polypyridine and Porphyrin complexes. Academic Press, London
2. Norris J, Meisel D (eds) (1989) Photochemical energy conversion. Elsevier, New-York
3. (a) Balzani V, Barigelleti F, De Cola L (1990) Topics in Current Chemistry 158: 33; (b) Demas JN, De Graff BA (1991) Anal. Chem. 63: 829A
4. Sykora J, Sima J (1990) Coord. Chem. Rev. 107: 1
5. Barton JK (1988) Chem. Eng. News September 26: 30
6. (a) Davidson RS, Hilchenbach MM (1990) Photochem. Photobiol. 52: 431. (b) For an example based on organic dyes, see Glazer AN, Hays SR (1992) Nature 359: 859
7. Jenkins Y, Barton JK (1992) J. Am. Chem. Soc. 114: 8736
8. Bannwarth W, Schmidt D, Stallard RL, Hornung C, Knorr R, Müller F (1988) Helvetica Chemica Acta 71: 2085
9. Telser J, Cruickshank K, Schanze KS, Netzel TL (1989) J. Am. Chem. Soc. 111: 7221; 7226
10. Basile LA, Barton JK (1989) In: Sigel H (ed) Metal ions in biological systems. Marcel Dekker, New York, p 32
11. Balzani, V, Scandola F (1991) Supramolecular photochemistry. Ellis Horwood, Chichester, UK
12. For reviews on phototherapy: (a) Henderson BW, Dougherty TJ (1992) Photochem. Photobiol. 55: 145; (b) Anderson-Engels S, Johansson J, Svanberg S, Svanberg K (1989) Anal. Chem. 61: 1367A
13. for example: Bruhn S, Toney J, Lippard SJ (1990) Progress in Inorganic Chemistry, Bioinorganic Chemistry 38: 477
14. Balzani V, Carassiti V (1970) Photochemistry of coordination compounds. Academic Press, London.
15. Adamson AW, Fleischauer PD (eds) (1975) Concepts of inorganic photochemistry. Wiley, New York.
16. Wrighton M (ed) (1978) Inorganic and organometallic photochemistry. ACS, Washington.
17. Wrighton M, Geoffroy GL (1979) Organometallic photochemistry. Academic Press, New York.

18. Ferraudi GJ (1988) Elements of Inorganic Photochemistry. Wiley, New York.
19. Hoffman MZ (1983) J. Chem. Ed. 60: 784 and subsequent papers.
20. Adamson AW (1967) J. Phys. Chem. 71: 798
21. Saenger W (1984) Principles in nucleic acid structure, Cantor CR (ed). Springer, Berlin, Heidelberg, New York
22. Kennard O (1984) Pure & Appl. Chem. 56: 989
23. Kennard O, Huntern W (1991) Angew. Chem. Int. Ed. Engl. 30: 1254
24. Dickerson RE (1992) In: Lilley DM, Dahlberg JE (eds) Methods in enzymology, vol 112: DNA structures. Part B. Chemical and electrophoretic analysis of DNA, Academic Press, New York
25. Dickerson RE, Drew HR, Conner BN, Wing RM, Fratini AV, Kopka ML (1982) Science 216: 475
26. Watson JD, Crick FH (1953) Nature 171: 737
27. Wang AH-J, Quigley GJ, Kolpak FJ, Crawford JL, van Boom JH, van der Marel G, Rich A (1979) Nature 282: 680
28. Drew HR, Takano T, Tanaka S, Itakura K, Dickerson RE (1980) Nature 286: 567
29. Sharp KA (1991) Current Opinion in Structural Biology 1: 171
30. Neidle S, Abraham Z (1984) CRC Crit. Rev. Biochem. 17: 73
31. Wang AH-J (1992) Current Opinion in Structural Biology 2: 361
32. Lerman LS (1961) J. Mol. Biol. 3: 18–30
33. Marzili LG (1977) Progress in Inorganic Chemistry 23: 255
34. Van Dyke MW, Hertzberg RP, Dervan PB (1982) Proc. Natl. Acad. Sci. USA 79: 5470
35. Kopka ML, Yoon C, Goodsell D, Pjura P, Dickerson RE (1985) Proc. Natl. Acad. Sci. USA 82: 1376
36. Zimmer C, Luck G, Thrum H, Pitra C (1972) Eur. J. Biochem. 26: 81
37. Zakrzewska K, Lavery R, Pullman B (1983) Nucl. Acids Res. 11: 8825
38. Long EC, Barton JK (1990) Acc. Chem. Res. 23: 271
39. Dougherty G, Pigram WJ (1982) CRC Crit. Rev. Biochem. 12: 103
40. Sherman SE, Lippard SJ (1987) Chem. Rev. 87: 1153
41. Sundquist WI, Lippard SJ (1990) Coord. Chem. Rev. 100: 293
42. Pui S, Frederik CA, Saal D, Wang AH-J, Rich A (1987) J. Biomolec. Struct. & Dynamics 4: 521
43. Satyanarayana S, Dabrowiak C, Chaires JB (1992) Biochemistry 31: 9319
44. Härd T, Hiort C, Nordén B (1987) J. Biomol. Struct. Dynam. 5: 89
45. (a) Scatchard G (1949) Ann. N.Y. Acad. Sci. 51: 660; (b) McGhee JD, von Hippel PH (1974) J. Mol. Biol. 86: 469
46. Zimmermann HW (1986) Angew. Chem. Int. Ed. Engl. 25: 115
47. Cohen G, Eisenberg H (1969) Biopolymers 8: 45
48. Barton JK, Danishefsky AT, Goldberg JM (1984) J. Am. Chem. Soc. 106: 2172
49. Kelly JM, Tossi AB, McConnell DJ, OhUigin (1985) Nucl. Acids Res. 13: 6017
50. Fisher LM, Kuroda R, Sakai TT (1985) Biochemistry 24: 3199
51. Barton JK, Goldberg JM, Kumar CV, Turro NJ (1986) J. Am. Chem. Soc. 108: 2081
52. Kumar CV, Barton JK, Turro NJ (1985) J. Am. Chem. Soc. 107: 5518
53. Yamagishi A (1983) J. Chem. Soc., Chem. Commun. 572
54. Yamagishi A (1984) J. Phys. Chem. 88: 5709
55. Hiort C, Nordén B, Rodger A (1990) J. Am. Chem. Soc. 112: 1971
56. Hiort C, Nordén B (1988) Nucleosides & Nucleotides 7: 661
57. Härd T, Hiort C, Nordén B (1987) J. Biomol. Struct. Dynam. 5: 89
58. Härd T, Nordén B (1986) Biopolymers 25: 1209
59. Shafer RH, Brown SC (1988) In: Kallenbach NR (ed) Chemistry and physics of DNA-ligand interactions, Adenine Press, Schenectady, NY, p 109
60. Nordén B, Patel N, Hiort C, Gräslund A, Kim SK (1991) Nucleosides & Nucleotides 10: 195
61. Eriksson M, Leijon M, Hiort C, Nordén B, Gräslund A (1992) J. Am. Chem. Soc. 114: 4933
62. Rehman JP, Barton JK (1990) Biochemistry 29: 1701
63. Rehman JP, Barton JK (1990) Biochemistry 29: 1710
64. Ottaviani MF, Ghatlia ND, Bossman SH, Barton JK, Dürr H, Turro NJ (1992) J. Am. Chem. Soc. 114: 8946
65. Pyle AM, Barton JK (1990) Prog. Inorg. Chem. 38: 413
66. Mei HY, Barton JK (1986) J. Am. Chem. Soc. 108: 7414
67. Mei HY, Barton JK (1988) Proc. Natl. Acad. Sci. USA 85: 1339

68. Görner H, Tossi AB, Stradowski C, Schulte-Frohlinde D (1988) J. Photochem. Photobiol., B: Biol. 2: 67
69. Friedman AE, Chambron JC, Sauvage JP, Turro NJ, Barton JK (1990) J. Am. Chem. Soc. 112: 4960
70. Friedman AE, Kumar CV, Turro NJ, Barton JK (1991) Nucl. Acids Res. 19: 2595
71. Hartshorn RM, Barton JK (1992) J. Am. Chem. Soc. 114: 5919
72. Hiort C, Lincoln P, Nordén B (1993) J. Am. Chem. Soc. 115: 3448
73. Kirsch-De Mesmaeker A, Orellana G, Barton JK, Turro NJ (1990) Photochem. Photobiol. 52: 461
74. Tossi AB, Kelly JM (1989) Photochem. Photobiol. 49: 545
75. de Buyl F, Kirsch-De Mesmaeker A, Tossi AB, Kelly JM (1991) J. Photochem. Photobiol. A: Chem. 60: 27
76. Kirsch-De Mesmaeker A, Jacquet L, Masschelein A, Vanhecke F, Heremans K (1989) Inorg. Chem. 28: 2465
77. Baker AD, Morgan RJ, Strekas TC (1991) J. Am. Chem. Soc. 113: 1411
78. Baker AD, Morgan RJ, Strekas TC (1992) J. Chem. Soc., Chem. Commun. 1099
79. Tysoe SA, Morgan RJ, Baker AD, Strekas TC (1993) J. Phys. Chem. 97: 1707
80. Morgan RJ, Chatterjee S, Baker AD, Strekas TC (1991) Inorg. Chem. 30: 2687
81. Orellana G, Kirsch-De Mesmaeker A, Barton JK, Turro NJ (1991) Photochem. Photobiol. 54: 499
82. Carter MT, Bard AJ (1990) Bioconjugate Chem. 1: 257
83. Kapicak L, Gabbay EJ (1975) J. Am. Chem. Soc. 97: 403
84. Haworth IS, Elcock AH, Freeman J, Rodger A, Richards WG (1991) J. Biomol. Structure & Dyn. 9: 23
85. Barton JK (1986) Science 233: 727
86. Juris A, Barigelletti F, Campagna S, Balzani V, Belzer P, von Zelewsky A (1988) Coord. Chem. Rev. 84: 85
87. Kalyanasundaram K (1982) Coord. Chem. Rev. 46: 159
88. Meyer TJ (1990) Pure Appl. Chem. 62: 1003
89. Van Houten J, Watts RJ (1976) J. Am. Chem. Soc. 98: 4853
90. Casper JV, Meyer TJ (1983) J. Am. Chem. Soc. 105: 5583
91. Jacquet L, Kirsch-De Mesmaeker A (1992) J. Chem. Soc. Faraday Trans. 88: 2471
92. Masschelein A, Jacquet L, Kirsch-De Mesmaeker A, Nasielski J (1990) Inorg. Chem. 29: 855
93. Crutchley RJ, Kress N, Lever ABP (1983) J. Am. Chem. Soc. 105: 1170
94. Kirsch-De Mesmaeker A, Jacquet L, Nasielski J (1988) Inorg. Chem. 27: 4451
95. Lecomte JP, Kirsch-De Mesmaeker A, Orellana G (1994) J. Phys. Chem. 98: 5382
96. Kelly JM, Feeney MM, Tossi AB, Lecomte JP, Kirsch-De Mesmaeker A (1990) Anti-Cancer Drug Design 5: 69
97. Kelly JM, Mc Connell DJ, OhUigin C, Tossi AB, Kirsch-De Mesmaeker A, Masschelein A, Nasielski J (1987) J. Chem. Soc. Chem. Comm. 1821
98. Neshvad G, Hoffman MZ, Mullazani Q, Venturi M, Ciano M, D'Angelantonio M, (1989) J. Phys. Chem. 93: 6080; D'angelantonio M, Mulazzani QG, Venturi M, Ciano M, Hoffman MZ (1991) J. Phys. Chem. 95: 5121
99. Masschelein A, Kirsch-De Mesmaeker A (1987) New J. Chem. 11: 329
100. Lecomte J-P, Kirsch-De Mesmaeker A, Kelly JM, Tossi AB, Görner H (1992) Photochem. Photobiol. 55: 681
101. Kittler L, Löber G, Gollmick F, Berg H (1980) J. Electroanal. Chem. 116: 503
102. Brabec V, Dryhurst G (1978) J. Electroanal. Chem. 89: 161
103. Jovanovic SV, Simic MG (1986) J. Phys. Chem. 90: 974
104. Values determined by pulse radiolysis: [$E°(G^{\cdot+}/G) = 1.33V$; $E°(A^{\cdot+}/A) > 1.75$ V vs NHE], Candeias L, Steenken S, Personal Communication
105. Aboul-Enein A, Schulte-Frohlinde D (1988) Photochem. Photobiol. 48: 27
106. Görner H, Stradowski C, Schulte- Frohlinde D (1988) Photochem. Photobiol. 47: 15
107. (a) Tossi AB, Görner H, Aboul-Enein A, Schulte-Frohlinde D (1989) Free Radical Res. Commun. 6: 171; (b) Tossi AB, Görner H, Schulte-Frohlinde D (1989) Photochem. Photobiol. 50: 585
108. Stradowski C, Görner H, Currel LJ, Schulte-Frohlinde D (1987) Biopolymers 26: 189
109. Bolleta F, Juris A, Maestra M, Sandrini D (1980) Inorg. Chim. Acta. 44: L175
110. (a) Whitten DG (1980) Acc. Chem. Res. 13: 83

110. (b) Tan-sien-Hee L, Kirsch-De mesmaeker A (1994) J. Photochem. Photobiol. A. in press
111. Fedorova OS, Podust LM (1988) J. Inorg. Biochem. 34: 149
112. (a) von Sonntag C, Schuchmann H-P. (1991) Angew. Chem. Int. Ed. Engl. 30: 1229; (b) Bielski BHJ, Cabelli DE (1991) Int. J. Radiat. Biol. 59: 2191
113. Barton JK, Kumar CV, Turro NJ (1986) J. Am. Chem. Soc. 108: 6391
114. Purugganan MD, Kumar CV, Turro NJ, Barton JK (1988) Science 241: 1645
115. Fromherz P, Rieger B (1986) J. Am. Chem. Soc. 108: 5361
116. Brun AM, Harriman A (1992) J. Am. Chem. Soc. 114: 3656
117. Fleisher MB, Waterman KC, Turro NJ, Barton JK (1986) Inorg. Chem. 25: 3549
118. Piette J (1991) J. Photochem. Photobiol., B: Biol. 11: 241
119. Kelly JM, Tossi AB, McConnell D, OhUigin C, Hélène C, Le Doan T (1989) Free Radicals, Metal Ions and Biopolymers, Richelieu
120. Cadet J, Vigny P (1990) In: Morrison H (ed) Bioorganic photochemistry, photochemistry and the Nucleic Acids, Vol. 1, John Wiley, NY, p 1
121. Steenken S (1989) Chem. Rev. 89: 503
122. Schulte-Frohlinde D, Simic MG, Görner H (1990) Photochem. Photobiol. 52: 1137
123. Von Sonntag C (1987) Chemical Basis of Radiation Biology. Taylor and Francis, London.
124. Feeney M, Kelly JM, Kirsch-De Mesmaeker A, Lecomte J-P, Tossi AB (1994) J. Photochem. Photobiol. B: Biol. 23: 69
125. Crosby GA, Elfing WH (1976) J. Phys. Chem. 80: 2206
126. Watts RJ, Van Houten J (1978) J. Am. Chem. Soc. 100: 1718
127. Nishizawa M, Suzuki TM, Watts RJ, Ford PC (1984) Inorg. Chem 23: 1837
128. Frink ME, Sprouse SD, Goodwin HA, Watts RJ, Ford PC (1988) J. Am. Chem. Soc. 27: 1283
129. Indelli MT, Carioli A, Scandola F (1984) J. Phys. Chem. 27: 2685
130. Pyle AM, Chiang MY, Barton JK (1990) Inorg. Chem. 29: 4487
131. Demas JN, Crosby GA (1970) J. Am. Chem. Soc. 92: 7262
132. Muir M, Huang W-L (1973) Inorg. Chem. 12: 1831
133. Ballardini R, Varani G, Balzani V (1980) J. Am. Chem. Soc. 102: 1719
134. Ohno T (1985) J. Phys. Chem. 89: 5709
135. Indelli M, Ballardini R, Scandola F (1984) J. Phys. Chem. 27: 1283
136. Sitlani A, Long EC, Pyle AM, Barton JK (1992) J. Am. Chem. Soc. 114: 2312
137. Gurzadyan GG, Görner H (1992) Photochem. Photobiol. 56: 371
138. Kirshenbaum MR, Tribolet R, Barton JK (1988) Nucl. Acids Res. 16: 7943
139. Pyle AM, Long EC, Barton JK (1989) J. Am. Chem. Soc. 111: 4520
140. Pyle AM, Morii T, Barton JK (1990) J. Am. Chem. Soc. 112: 9432
141. Chow CS, Barton JK (1990) J. Am. Chem. Soc. 112: 2839
142. Mahnken RE, Billadeau MA, Nikonowicz EP, Morrisson H (1992) J. Am. Chem. Soc. 114: 9253
143. Mahnken RE, Bina M, Deibel RM, Morrison H (1989) Photochem. Photobiol. 49: 519
144. Balzani V, Carassiti V (1970) Photochemistry of coordination compounds. Academic Press, London, p 193
145. Adamson AW (1968) Coord. Chem. Rev. 3: 169
146. Martin B, Waind GM (1958) J. Chem. Soc. 4284
147. Barton JK, Raphael AL (1984) J. Am. Chem. Soc. 106: 2466
148. Barton JK, Raphael AL (1985) Proc. Natl. Acad. Sci. USA 82: 6460
149. Müller BC, Raphael AL, Barton JK (1987) Proc. Natl. Acad. Sci. USA 84: 1764
150. Chapnick LB, Chasin LA, Raphael AL, Barton JK (1988) Mut. Res. 201: 17
151. Saito I, Morii T, Sugiyama H, Matsuura T, Meares CF, Hecht SM (1989) J. Am. Chem. Soc. 111: 2307
152. Farinas E, Tan JD, Baidya N, Mascharak PK (1993) J. Am. Chem. Soc. 115: 2996
153. Tan JD, Hudson SE, Brown SJ, Olmstead MM, Mascharak PK (1992) J. Am. Chem. Soc 114: 3853
154. Kvam E, Moan J (1990) Photochem. Photobiol. 52: 769.
155. Fiel RJ (1989) J. Biomol. Struct. Dynamics 6: 1259
156. Marzilli LG (1990) New J. Chem 14: 409
157. Gibbs EJ, Pasternack RF (1989) Seminars Haemat 26: 77
158. Sari MA, Battioni JP, Dupré D, Mansuy D, Le Pecq JB (1990) Biochemistry 29: 4205
159. Kalyanasundaram K (1992) Photochemistry of polypyridine and porphyrin complexes. Academic Press, London, p 369

160. Kelly JM, Murphy MJ, McConnell DJ, OhUigin C (1985) Nucl. Acids Res 13: 167
161. Murphy MJ (1986) PhD thesis, University of Dublin, Ireland
162. Liu Y, Konigstein JA, Evdokimov YM (1991) Can. J. Chem. 69: 1791.
163. Le Doan T, Perrouault L, Rougee M, Bensasson RV, Helene C (1985) In: Jori G, Perria C (eds) Photodynamic Therapy of Tumours and other Diseases, Libreria Progetto, Padova, p 56
164. Sari MA, Battioni JP, Mansuy D, Le Pecq JB (1986), Biochem. Biophys. Res. Comm. 141: 643
165. Pasternack RF, Caccam M, Keogh B, Stephenson TA, Williams AP, Gibbs EJ (1991) J. Am. Chem. Soc. 113: 6835.
166. (a) Milder SJ, Ding L, Etemad-Moghadam G, Meunier B, Paillous N (1990) J. Chem. Soc., Chem. Commun. 1131.
166. (b) Sentagne C, Meunier B, Paillous N (1992) J. Photochem. Photobiol. B. 16: 47
167. Turpin PY, Chinsky L, Laigle A, Tsuboi M, Kincaid JR, Nakamoto K (1989) Photochem. Photobiol. 51: 519.
168. Chinsky L, Turpin PY, Al-Obaidi AHR, Bell SEJ, Hester RE (1991) J. Phys. Chem. 95: 5754
169. Hudson BP, Sou J, Berger DJ, McMillin DR (1992) J. Am. Chem. Soc. 114: 8997
170. Le Doan T, Perrouault L, Helene C, Chassignol M, Thuong NT (1986) Biochemistry 25: 6739
171. Pitié M, Pratviel G, Bernadou J, Meunier B (1992) Proc. Natl. Acad. Sci. 89: 3967
172. Meunier B (1992) Chem. Rev. 92: 1411
173. (a) Fiel RJ, Datta-Gupta N, Mark EH, Howard JC (1981) Cancer Res. 41: 3543; (b) Munson BR, Fiel RJ (1992) Nucl. Acids Res. 20: 1315
174. Praseuth D, Gaudemer A, Verlhac J-B, Kraljic I, Sissoeff I, Guillé E (1986) Photochem. Photobiol. 44: 717.
175. Croke DT, Perrouault L, Sari MA, Battioni J-P, Mansuy D, Helene C, Le Doan T (1993) J. Photochem. Photobiol. B: Biol. 18: 41.
176. Le Doan T, Praseuth D, Perouault L, Chassignol M, Thuong NT, Hélène C (1990) Bioconjugate Chem. 2: 108.
177. (a) Shoikhet KG, Kazantsev AV, Federova OS (1991) Nucl. Acids Res. (symposium Series) 24: 248; (b) Englisch U, Gauss DH (1991) Angew. Chem. Int. Ed. Engl. 30: 613
178. Villanueva A, Canete M, Juarranz A, Stockert JC (1987) Basic Appl. Histochem. 31: 9.
179. Villanueva A, Canete M, Hazen MJ (1989) Mutagenesis 4: 157.
180. Rück A, Köllner T, Dietrich A, Strauss W, Schneckenburger H (1992) J. Photochem. Photobiol. B: Biol. 12: 403.
181. Kasturi C, Platz MS (1992) Photochem. Photobiol. 56: 427.
182. Burrows HD, Kemp TJ (1974) Chem. Soc. Rev. 3: 138
183. Azenha MEDG, Burros HD, Formoshino SJ, Miguel MGM (1989) J. Chem. Soc., Faraday Trans. 85: 2625
184. Cunningham J, Srijaranai S (1991) J. Photochem. Photobiol. A: Chem. 55: 219.
185. Nielsen PE, Jeppesen C, Buchardt O (1988) FEBS Lett. 235: 122
186. Nielsen PE, Hiort C, Sonnichsen SH, Buchardt O, Dahl O, Norden B (1992) J. Am. Chem. Soc. 114: 4967
187. Jeppesen C, Nielsen PE (1989) Nucl. Acids Res. 17: 4947
188. Nielsen PE, Jeppesen C (1990) Trends Photochem. Photobiol. 1: 39
189. Nielsen PE, Mollegaard NE, Jeppesen C (1990) Nucl. Acids Res. 18: 3847
190. Nielsen PE (1992) Nucl. Acids. Res. 20: 2735
191. Ahorthe I, Elcock A, Rodger A, Richards W (1991) J. Biomol. Struct. Dynam. 9: 553
192. Barton JK, Paranawithana SR (1986) Biochemistry 25: 2205
193. Pyle AM, Rehmann JP, Meshoyrer R, Kumar CV, Turro NJ, Barton JK (1989) J. Am. Chem. Soc. 111: 3051
194. Naing K, Takahashi M, Taniguchi M, Yamagashi A (1993) J. Chem. Soc. Chem. Commun. 402
195. Kojima H, Sato N, Kawamoto Y, Iyoda J (1989) Chem. Let. 1889
196. Barton JK, Basile LA, Danishefsky A, Alexandrescu A (1984) Proc. Natl. Acad. Sci. USA 81: 1961
197. Basile LA, Barton JK, (1987) J. Am. Chem. Soc. 109: 7548
198. Basile LA, Raphael AL, Barton JK (1987) J. Am. Chem. Soc. 109: 7550
199. Barton JK, Lolis E (1985) J. Am. Chem. Soc. 107: 708
200. Carter MT, Rodriguez M, Bard AJ (1989) J. Am. Chem. Soc. 111: 8901
201. David SS, Barton JK (1993) J. Am. Chem. Soc. 115: 2984
202. Uchida K, Pyle AM, Morii T, Barton JK (1989) Nucl. Acids Res. 17: 10259
203. Grover N, Thorp HH (1991) J. Am. Chem. Soc. 113: 7030
204. Grover N, Gupta N, Thorp HH (1992) J. Am. Chem. Soc. 114: 3390

205. Grover N, Gupta N, Singh P, Thorp HH (1992) Inorg. Chem. 31: 2014
206. Barton JK, Dannenberg JJ, Raphael AL (1982) J. Am. Chem. Soc. 104: 4967
207. Gupta N, Grover N, Neyhart GA, Liang W, Singh P, Thorp HH (1992) Angew. Chem. Int. Ed. Engl. 31: 1048
208. Carter MT, Bard AJ (1987) J. Am. Chem. Soc. 109: 7258
209. Gupta N, Grover N, Neyhart GA, Singh P, Thorp HH (1993) Inorg. Chem. 32: 310
210. Norden B, Tjerneld F (1976) FEBS Lett. 67: 368
211. Lecomte J-P, Kirsch-De Mesmaeker A, Demeunynck M, Lhomme J (1993) J. Chem. Soc. Faraday Trans. 89: 3261
212. Mestroni G, Alession E, Calligaris M, Attia WM, Quadrifoglio F, Cauci S, Sava G, Zorzet S, Pacor S, Monti-Bragadin C, Tamaro M, Dolzani L (1989) 10: 71

Received: March 1994

Radical Ion Cyclizations

Sandra Hintz[1], Andreas Heidbreder[2] and Jochen Mattay*[1]

Organisch-Chemisches Institut der Universität Münster, Corrensstr. 40, D-48149 Münster, Germany
[1] New address: Institut für Organische Chemie der Universität Kiel, Olshausenstr. 40 (Otto-Hahn-Platz 4), D-24098 Kiel, Germany
[2] New address: Henkel KGaA, D-40191 Düsseldorf, Germany

Table of Contents

Topics in Current Chemistry, Vol. 177

Radical ions generated by single-electron transfer from neutral organic compounds are known as important intermediates in a variety of interesting chemical processes and reactions. Our aim in this article will be to provide a comprehensive review of the design and application of cyclization reactions that utilize radical ions as reactive intermediates. Examples from the last ten years will be reported and mechanistic aspects will be discussed critically. The cyclization reactions are classified into three categories regardless of the method of generating of the reactive species. After a brief introduction, cyclization reactions via radical cations are described. The subsequent category is devoted to radical-anion-mediated cyclization reactions. The last category includes cyclization reactions via reactive intermediates containing both radical anions and radical cations.

1 Introduction

The construction of complex carbocyclic and heterocyclic ring systems in a regio- and stereoselective fashion remains one of the fundamental problems in synthetic organic chemistry. Beside pericyclic and metal-catalyzed reactions many strategies have been developed involving anionic or cationic intermediates [1, 2]. In recent years, free-radical cyclization reactions have been established as valuable synthetic tools in organic synthesis [3–7]. Although free radicals are known to be highly reactive species, their intramolecular addition to carbon-carbon multiple bonds or even to carbon-nitrogen or carbon-oxygen multiple bonds proceeds with high, predictable selectivity [8–10]. Because of the chain nature of free radical reactions, many possibilities are offered to control the course of the cyclization reactions.

Whereas the design and application of free radical cyclization reactions have been extensively covered in excellent reviews [3–5], there is no comprehensive report on the synthetic application of their charged counterparts: radical cations and radical anions.

Radical ions – charged species with unpaired electrons – are easily generated by a number of methods that are discussed in more detail below. Their properties have been characterized by several spectroscopic techniques, and their structures and spin density contributions have been the subject of molecular orbital calculations at different levels of sophistication. The behaviour of radical ions in rearrangement and isomerization reactions as well as in bond-cleavage reactions has been extensively studied [for recent reviews see Refs. 11–13 and references cited therein]. Useful synthetic applications, such as the radical-cation-catalyzed cycloaddition [14–20] or the *anti*-Markovnikov addition of nucleophiles to alkenyl radical cations [21–25], have been well documented. In analogy, the addition of electrophiles to alkenyl radical anions in a Markovnikov fashion has been intensively studied [24, 26]. Perhaps the largest class of radical anion fragmentation reactions is represented by the aromatic nuleophilic substitution via the $S_{RN}1$ mechanism. In general terms the $S_{RN}1$ mechanism is a chain reaction with single-electron transfer as the initial step [27–31]. In this respect radical anion reactions of nitro compounds have also been extensively studied. The nitro group is a good electron acceptor because of a low-energy π^* molecular orbital which allows the formation of a relatively stable radical anion. The most common course of the reaction is dissociation of the radical anion to a radical and an anion, followed by further reactions of the radical [32–34].

This article is intended to review recent reports on cyclization reactions that utilize radical anions or radical cations as reactive intermediates. The main characteristic of the term "cyclization reaction" is thought to be the intramolecular interaction of one radical-ion-activated site of a bifunctional precursor with a second non-activated site – e.g. a carbon-carbon or a carbon-heteroatom multiple bond – to form a cyclic product. Pericyclic reactions (cycloadditions, electrocyclic reactions, and sigmatropic rearrangements) of radical ions will not be included. After briefly outlining some common methods for generating radical ions, the authors wish to discuss critically their application to synthetic organic chemistry, covering literature reports that start at the beginning of the past decade.

Primarily, the common methods for generating organic radical cations starting from neutral compounds are based on one of the following processes [11, 35, 36]: chemical single-electron transfer oxidation [37–42], anodic oxidation [43–49], or photoinduced electron transfer (PET) oxidation [16, 50–53]. Other important methods mainly applied to analytical purposes include the radiolytical generation of radical cations [54] and the generation of molecular ions by electron impact ionization [55, 56]. The formation of a radical cation from a neutral organic donor by electron transfer oxidation may be generalized as follows (Eq. 1):

$$RH \xrightarrow{-e^-} RH^{+\bullet} \tag{1}$$

The feasibility of electron transfer oxidation is dictated by the thermodynamic potential E°_{ox} of the substrate RH and requires an anode potential or an oxidant to match the value of E°_{ox}. It is essential to choose an oxidant with an one-electron reduction potential sufficient for the desired oxidation and a two-electron reduction potential insufficient for further oxidation of the radical cation. The suitable oxidant may be a metal ion, a stable radical cation, or a typical PET-acceptor in its excited state. The advantage of electrochemically performed oxidation is obvious.

Due to the fact that the removal of a bonding electron from the HOMO of the substrate RH leads to a radical cation with enhanced reactivity with respect to fragmentation reactions, the pathway often employed in radical cation chemistry results in the separation of charge and spin by dissociative processes, such as deprotonation, desilylation or cleavage of a stable cationic leaving group

(Eq. 2). The free radical intermediates obtained frequently undergo second single-electron transfer oxidation to yield carbenium ions.

$$RH^{+\bullet} \xrightarrow{-H^+} R^{\bullet} \xrightarrow{-e^-} R^+ \qquad (2)$$

Again, it seems to be fundamental to select the suitable oxidant or to take advantage of electrochemical methods. Moreover, the deprotonation of radical cations can be controlled by conducting the oxidation reactions in buffered media.

On the one hand, this particular feature makes it more difficult to distinguish between reactions involving radical cations, free radicals or carbenium ions, but on the other hand the chemist acquires an additional tool to control the course of the intended reaction. Some illustrative examples of cyclization reactions that utilize cleavage of the radical cations, primarily generated by single-electron oxidation, will be given in the following sections.

The microscopic reverse of Eq. (2), the protonation of free radicals (such as, for example, aminium radicals) offers an additional access to aminium radical cations [57–60] (Eq. 3).

$$R^{\bullet} \xrightarrow{+H^+} RH^{+\bullet} \qquad (3)$$

Electron attachment to the LUMO of a neutral organic acceptor produces a radical anion [61]. This process can be initiated either chemically using a one-electron reducing agent [62, 63], electrochemically by cathodic reduction [64, 65] or photochemically in the presence of an electron donor in its excited state [12, 66].

$$RH \xrightarrow{+e^-} R^{-\bullet} \qquad (4)$$

Since there are only a few one-electron reducing agents (such as, for example, lithium 2,4,6-tri-*tert*-butylnitrobenzenide or potassium 1-(*N*,*N*-dimethylamino)-naphthalenide) chemical reduction plays a subordinated role in radical anion cyclization reactions. Depending on the reaction conditions and the starting material, alkali-metal reduction does not necessarily occur via radical anions. Single-electron transfer might be followed by a second electron transfer, creating dianions [67]. In several mechanistic studies, the reaction pathway of alkali-metal reduction has been elucidated [62, 68–70]. Initially formed radical anions are often involved in cyclization reactions but are not considered as reactive intermediates. Subsequent protonation or loss of an anionic leaving group leads to free radicals which undergo cyclization [71–73].

2 Radical Cation Cyclizations

Following our initial definition of the term "cyclization reaction", the starting materials possess one donor site which is easily oxidized by one of the above-mentioned methods and a second moiety attached at a suitable distance which is more stable towards the oxidizing agent. This moiety acts as the target of an intramolecular attack of the resulting radical cation. The electron donors generally employed are typical π-, σ-, or n-donors, such as alkenes, alkynes, aromatic hydrocarbons or amines. Since radical cations combine both spin and charge, they may react in cyclization reactions either like cations or like free radicals. Therefore, the functional group which is attacked by the radical cation moiety is most commonly represented by a multiple-bond-system, but even cyclizations to moieties providing active hydrogen are frequently used.

For clarity and convenience the following examples of radical-ion cyclization reactions are compiled according to the type of the electron donor from which the radical cation is generated.

2.1 Alkenyl Radical Cations

Similar to the intramolecular addition of neutral carbon-centered radicals to alkenes, the formation of radical cations starting from alkenes with subsequent cyclization offers a convenient method for constructing carbocyclic ring systems. In contrast to the regioselective 1,5-ring closure (5-*exo*-trig cyclization) of the 5-hexenyl radical [8, 9], the analogous α,ω-diene radical cation cyclizes in a 6-*endo*-trig mode. This generally observed preference for *endo*-cyclization is one important feature of radical cation cyclization reactions and is often quoted to distinguish between pathways involving radical cations or free radicals as reactive intermediates. The simplest case, the conversion of 1,5-hexadiene **1** to the 5-hexenyl radical cation **2** has been examined by Williams and co-workers in an ESR spectroscopical study [74]. The α,ω-diene radical cation **2** generated by γ-irradiation of diluted solid solutions of 1,5-hexadiene **1** reacts by *endo*-cyclization to form the cyclohexene radical cation **4**. The cyclohexane-1,4-diyl radical cation in its chair **3a** or boat **3b** form has been detected as an intermediate species which is converted to the product **4** by a photoinduced or thermally promoted 1,3-hydrogen shift, or by two consecutive 1,2-hydrogen shifts (Scheme 1).

Since the single-electron oxidation of electron-rich olefins, such as enols, enol ethers, enol acetates, or ketene acetals, is thermodynamically favored compared to simple alkenes, a number of attempts have been made to use

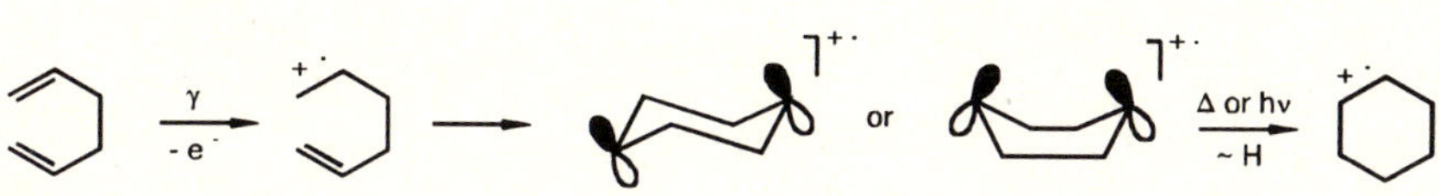

1 2 3a 3b 4

Scheme 1

alkenyl-substituted derivatives of these substrates as starting materials in cyclization reactions. Moreover, the selective oxidation of the more electron-rich double bond is easy to perform by a suitable oxidant and permits the cyclization of even unsymmetrical dienes in a predictable fashion. The anodic oxidation of 1-acetoxy-1,6-heptadiene homologues **5** in acetic acid to give mainly cyclohexenyl ketones **7** by intramolecular cyclization was first reported by Shono et al. in 1978 [75]. As byproducts, ketones **8** are obtained by hydrolysis of the starting enol acetates. In some cases considerable amounts of α,β-unsaturated ketones are formed by trapping of the radical cationic intermediate **6** with acetic acid.

Again, the exclusive formation of six-membered rings indicates that the cyclization takes place by the electrophilic attack of a cationic center, generated from the enol ester moiety to the olefinic double bond. The eventually conceivable oxidation of the terminal double bond seems to be negligible under the reaction conditions since the halve-wave oxidation potentials $E_{1/2}$ of enol acetates are + 1.44 to + 2.09 V vs. SCE in acetonitrile while those of 1-alkenes are + 2.70 to + 2.90 V vs. Ag/0.01 N $AgClO_4$ in acetonitrile and the cyclization reactions are carried out at anodic potentials of mainly 1.8 to 2.0 V vs. SCE.

Since enol silyl ethers are readily accessible by a number of methods in a regioselective manner and since the trialkylsilyl moiety as a potential cationic leaving group facilitates the termination of a cyclization sequence, unsaturated 1-trialkylsilyloxy-1-alkenes represent very promising substrates for radical-cation cyclization reactions. Several methods have been reported on the synthesis of 1,4-diketones by intermolecular oxidative coupling of enol silyl ethers with Cu(II) [76, 77], Ce(IV) [78], Pb(IV) [79], Ag(I) [80] V(V) [81] or iodosobenzene/BF_3-etherate [82] as oxidants without further oxidation of the products.

Snider and Kwon use either cupric triflate and cuprous oxide or ceric ammonium nitrate and sodium bicarbonate as single-electron oxidants to convert δ,ε- and ε,ζ-unsaturated enol silyl ethers **9** stereoselectively to the tricyclic ketones **14** in excellent yields [83, 84]. Based on comparison with other experimental data and literature results, the authors try to distinguish between several possible intermediates and propose the following mechanism with a very electrophilic radical cation **10** as the key intermediate.

After initial one-electron oxidation of **9** to the radical cation **10**, intramolecular addition to the olefinic double bond takes place to yield the cyclic radical

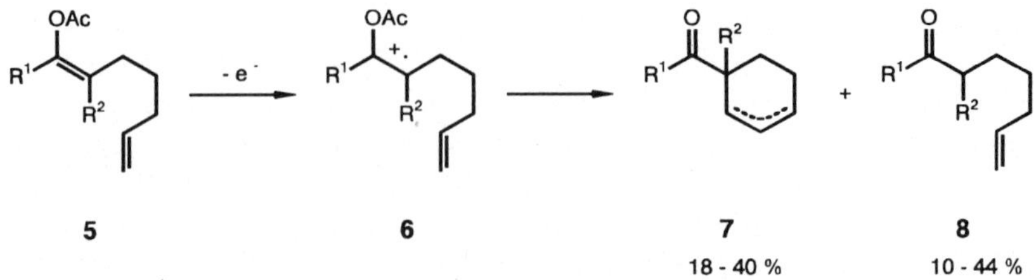

Scheme 2

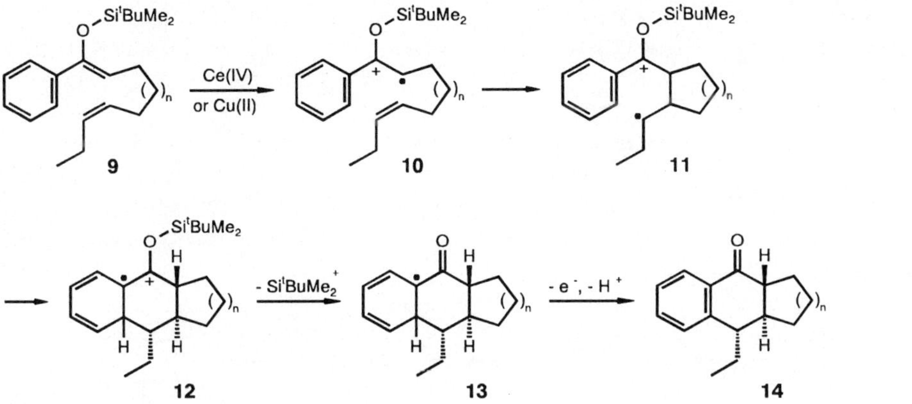

Scheme 3

cation **11**. Cyclization of the radical cation **11** to the benzene ring gives the radical cation **12** which loses the trialkylsilyl group to form the free radical **13**. Oxidation of the radical **13** and deprotonation finally yields the α-tetralone **14** (Scheme 3).

It is noteworthy that similar products are also obtained from oxidation of enol silyl ethers with two equivalents of tris(*p*-bromo-phenyl)aminium hexachloroantimonate which is known to oxidize electron-rich double bonds to cation radicals [15].

While simple trimethylsilyl enol ethers are not stable under the reaction conditions the oxidative cyclization of the more hydrolytically stable *tert*-butyldimethylsilyl enol ethers has proved to be applicable to a variety of starting materials including alkynyl substituted derivatives. Attempts to trap intermediates analogous to **11** by intramolecular addition to a second olefinic side chain failed since the second cyclization occurs exclusively to the benzene ring. An alternative method for the cyclization of radical cations obtained from enol silyl ethers, which requires only catalytical amounts of an oxidant, was recently reported by Heidbreder and Mattay [85]. Irradiation of 1-silyloxy-1,7-heptadienes **15** in the presence of 0.1 equivalent of the PET-sensitizer 9,10-dicyanoanthracene (DCA) leads to cyclic ketones **18**. The exclusive formation of 6-*endo* products gives rise to the assumption of radical cations **16** as key intermediates since cyclization via an α-carbonyl radical would be expected to yield a 3:1 mixture of 5-*exo* and 6- *endo* products [86]. Cyclization via an α-carbonyl cation is unlikely since cyclization of the much more stable *tert*-butyldimethylsilyl derivative leads to the same product in nearly the same yield as the reaction of the trimethylsilyl enol ether. Furthermore this methodology has successfully been applied to the regio- and stereoselective synthesis of bicyclic ketones **22**. As starting materials, easily accessible monocyclic enol silyl ethers **19**, tethered to a side-chain double bond at a suitable distance, are used.

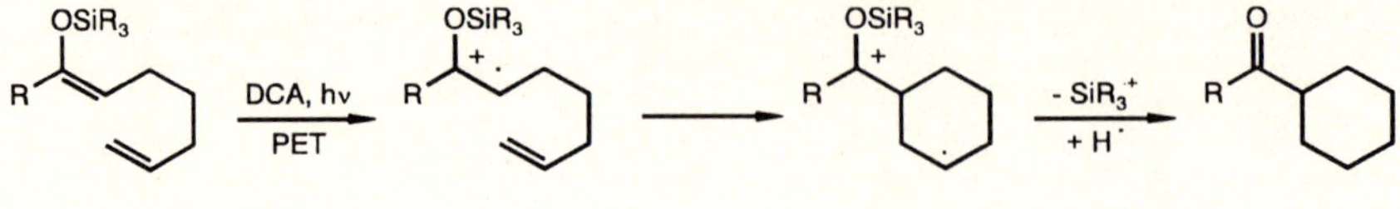

15 16 17 18

Scheme 4

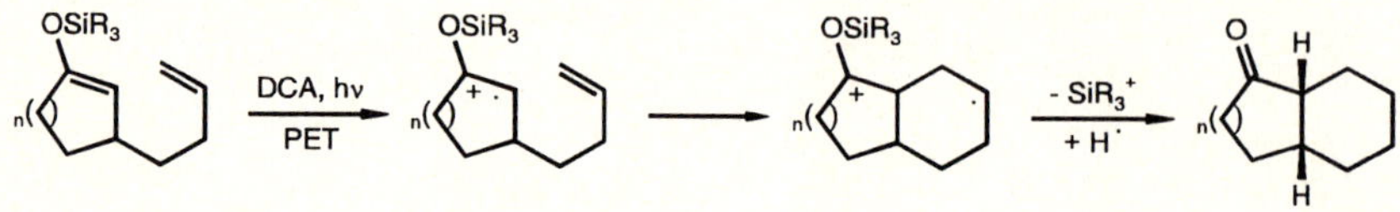

19 20 21 22

n=1,2

Scheme 5

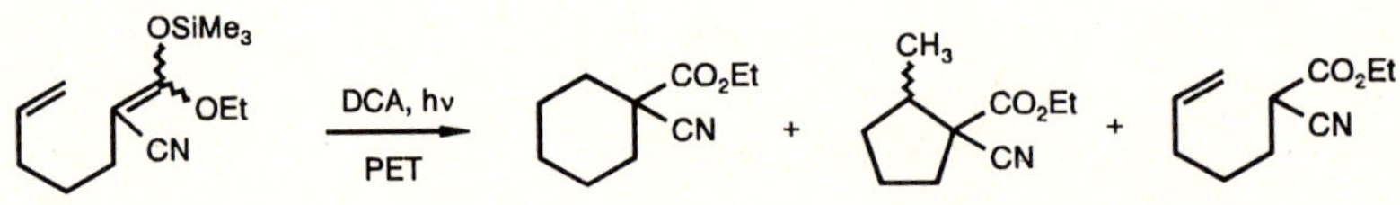

23 24 25 26

26 % <5 % 26 %

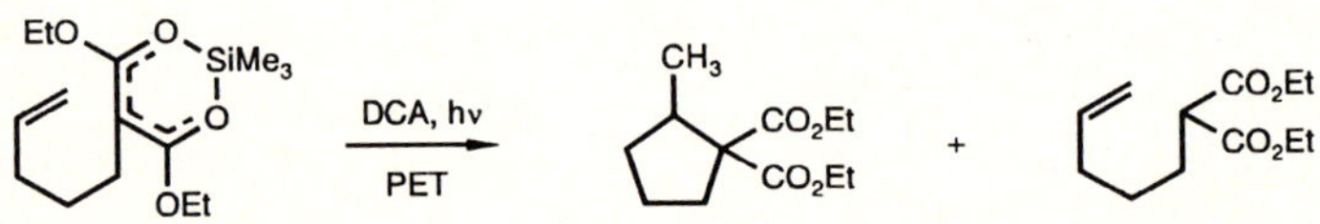

27 28 29

29 % 70 %

Scheme 6

Irradiation of the trimethylsilyl derivative as well as the *tert*-butyldimethylsilyl derivative under previously described conditions exclusively yields 6-*endo* products with a *cis*-ring juncture [87, 88] (Scheme 5).

The PET-oxidative cyclization of unsaturated O-alkyl-O-trimethylsilyl ketene acetals **23** and **27** yields cyclic esters **24**, **25**, and **28**, accompanied by the formation of considerable amounts of non-cyclic esters **26** and **29**, respectively [89]. The cyclization mode is found to be in accordance with free radical cyclizations of the appropriate esters **26** and **29**, performed by heating with organic peroxides [90]. Since organic electrochemistry can be used to oxidize electron-rich olefins, anodic oxidation seems to be useful for initiating intramolecular carbon-carbon bond formation of acid-sensitive unsaturated enol ethers. Moeller et al. discovered the formation of dimethoxy acetal products **31** and **32** after electrolysing enol silyl ether **30** under constant current conditions in an undivided cell using a 1 N $LiClO_4$ in 50% methanol-tetrahydrofuran solution as the electrolyte [91, 92]. The intramolecular anodic coupling of alkyl enol ethers **33** proceeds in comparably high yields. Again, cyclization of a radical cation precursor **34**, subsequent oxidation of the resulting secondary radical **35** and nucleophilic addition of methanol yields the bis-dimethoxy acetal **36** (Scheme 8).

In all of the cyclization reactions, Moeller has found only a small difference between the use of alkyl and silyl enol ethers. Since both styrenes and enol ethers have similar oxidation potentials, even the styrene moiety could function as the initiator for oxidative cyclization reactions. The anodic oxidation of simple styrene type precursors leads to low yields of cyclized products so that enol ether moiety seems to be the more efficient initiator for intramolecular anodic coupling reactions [93].

Moeller et al. have recently investigated intramolecular anodic olefin coupling reactions involving allyl- and vinylsilane groups [94–96]. Allylsilanes **37** exclusively provide *exo*-cyclized ring products, whereas the vinylsilane **40** predominantly leads to *endo*-cyclized six-membered ring products in analogy to the oxidation of the enol acetates shown by Shono et al. [75]. The enolether **43** containing an electron rich aromatic ring is also cyclized by anodic olefin coupling providing the carboanellated products **44** and **45** [97] (Scheme 10). Due to the similar oxidation potentials $E_{1/2}$ vs. Ag/AgCl (+ 1.35 V of **43**, + 1.53 V and + 1.61 V of the regioisomers **44**), it is conceivable that **45** arose from overoxidation of **44**. This is confirmed by the product ratio 2.7:1 (**44**:**45**)

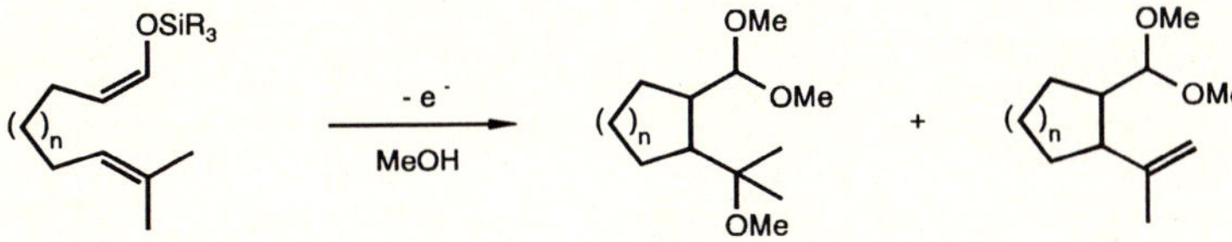

30 **31** **32**

Scheme 7

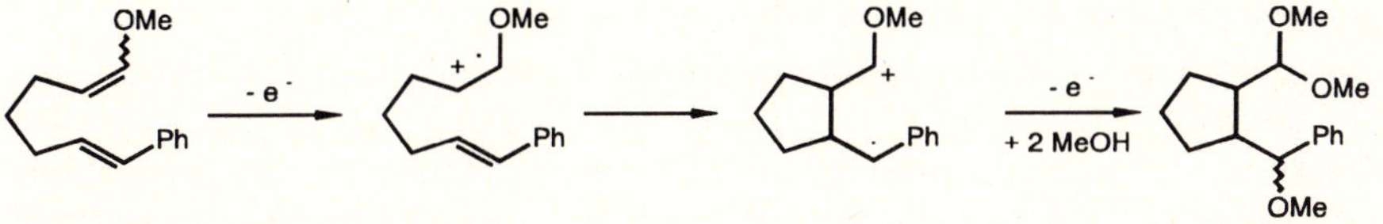

33 **34** **35** **36**

Scheme 8

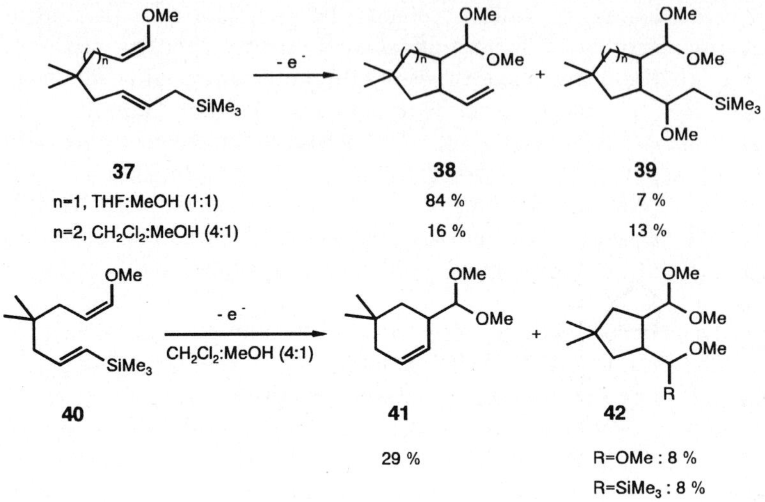

Scheme 9

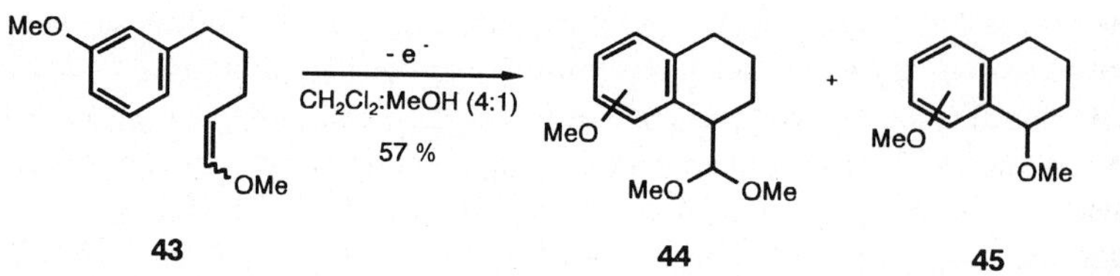

Scheme 10

using constant current electrolysis and 16:1 (**44**:**45**) using a controlled potential of 1.10 V. Analogous cyclization reactions are feasible with subtrates containing furan or pyrrole rings instead of homonuclear rings, yielding bicyclic products up to 75%.

Besides previously discussed enol ethers, enamines represent a potential radical cation precursor due to their electron-rich double bond. Cossy has recently investigated the oxidation of *N*-alkyl-*N*-unsaturated alkyl-β-carboxamindoenamines **46** derived from benzylamine or pyrrolidine by various metallic salts, such as AgOAc, $Co(OAc)_2$ and $Mn(OAc)_3$. The supposed radical cation intermediate **47** adds efficiently to the unactivated double bond producing spirolactams **49** and non-hydrolized iminolactams **50** [98, 99]. Generally, enamines obtained from benzylamine afford the better yields with less diastereoselectivity compared to analogous pyrrolidine-derived enamines. The latter show a diastereoselectivity ratio for the spirolactamization of about 95:5. The formation of the major diastereomer is explained by steric reasons, because the conformer that places a hydrogen atom rather than a vinyl group on top of the rigid pyrrolidino moiety is able to cyclize faster.

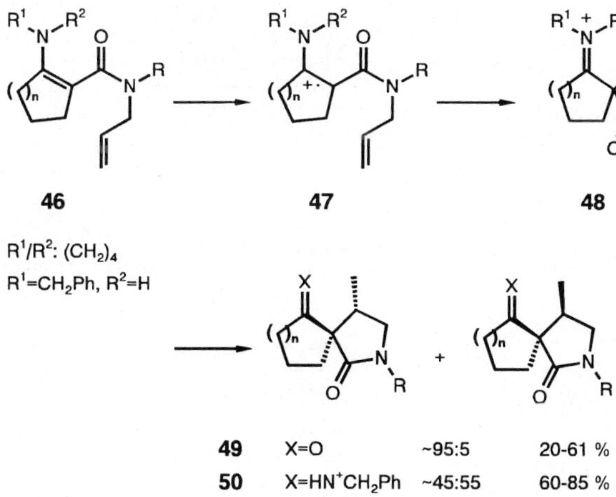

Scheme 11

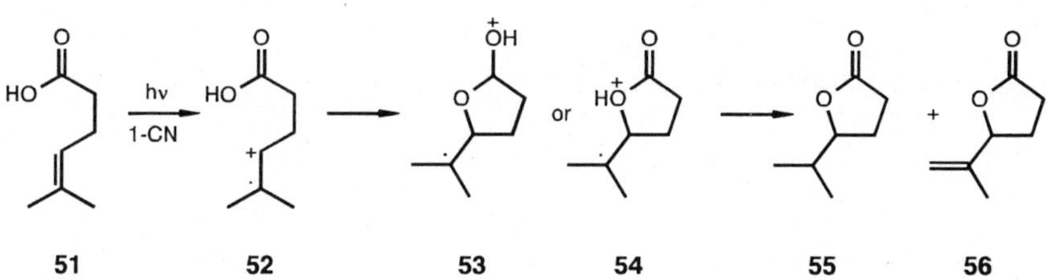

Scheme 12

Gassman and co-workers have recently investigated the photoinduced cyclization of γ,δ-unsaturated carboxylic acids to γ-lactones [100]. 5-Methyl-4-hexenoic acid **51** is converted into the radical cation **52**. The cyclization of **52** is expected to yield the distonic radical cations **53** and **54**, which leads to the *anti*-Markonikow products **55** and **56** in ratio 5:1. Fine-tuning of the sensitizer improved combined yields up to 89% [101].

Another heterocyclization is presented by Panifilow et al. Cyclic acetals and ethers are obtained by electrochemical oxidation of the terpenoid alcohol linalool **57** in methanol containing alkaline and sodium methoxide as electrolyt [102]. Anodic oxidation of the C(6)–C(7) double bond of linalool leads to the radical cation **58**. In addition to direct methoxylation of the radical cation an attack on the hydroxyl group takes place. After a second one-electron oxidation and following methoxylation the regioisomeric cyclic acetal and a subsequent 1,2-hydride shift, the cyclic acetal **60** and the cyclic ether **61** are finally formed in yields of 16 and 24%, respectively (Scheme 13). As shown by Utley and co-workers bicyclic lactones **65** and **66** can be synthesized by anodic oxidation

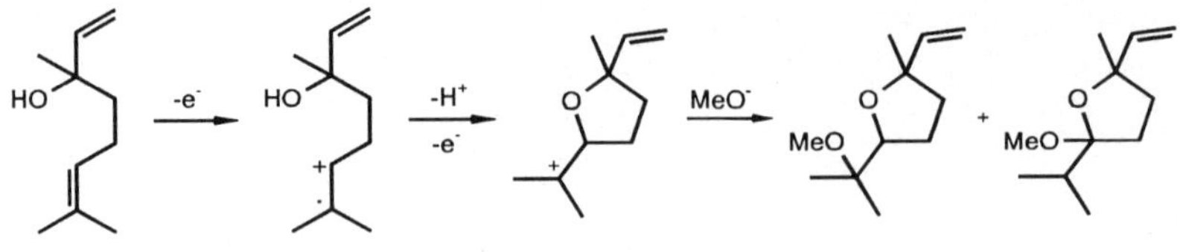

57 58 59 60 61

16 % 24 %

Scheme 13

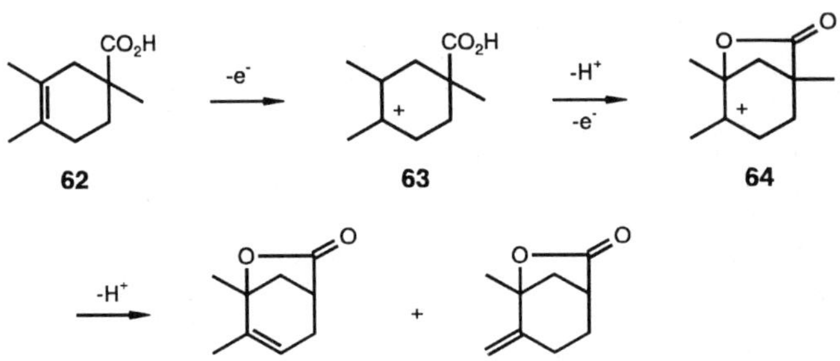

62 63 64

65 66

44 % 38 %

Scheme 14

of 1,2,4- trimethylcyclohexene-4-carboxylic acid **62** (Scheme 14). Since the oxidation potential of the double bond is sufficiently lowered, it is more readily oxidized than the carboxylate function [103].

Minisci and co-workers have investigated the oxidation of 4-penten-1-ol **67** by peroxydisulfate, known as one of the strongest oxidizing agents. Thermal, photochemical, radiolytic or redox decomposition of the peroxydisulfate ion supplies the radical anion $SO_4^{-\bullet}$ which appears to be a very effective electron-transfer oxidizing agent [104, 105]. Competition for oxidation between the hydroxyl group and the olefinic bond characterizes reactions of such olefinic alcohols [106]. The intermediate radicals **70** and **71** can either arise from oxidation of the hydroxyl group and intramolecular addition to the C–C-double bond or from oxidation of the double bond and subsequent nucleophilic attack of the alcohol function. By comparison with intramolecular addition of alkoxy radicals exclusively leading to 5-*exo* products [107], different regioselectivity is observed and therefore the second reaction pathway is more likely. The authors consider an electron transfer from the olefin and the intermediate formation of a radical cation as probable. Presumably the ionization potential of the olefin, in comparison with those of the hydroxy group, plays an important role in determining the isomeric ratio.

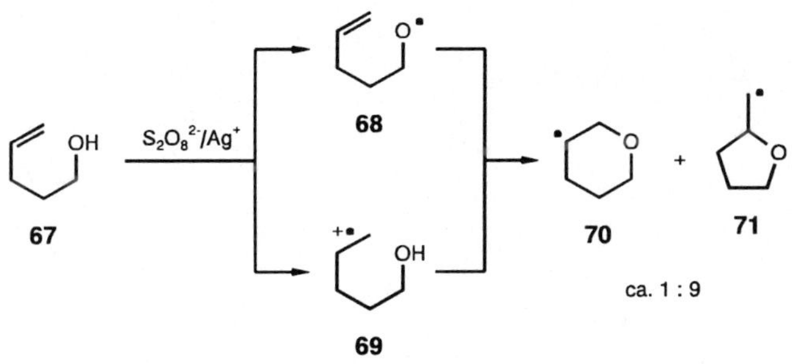

Scheme 15

The one-electron chemistry of enols has been intensively studied by Schmittel [108]. He has shown that the thermodynamic stability order of the ketone tautomer and the enol tautomer in the solution phase is inverted upon one-electron oxidation [109, 110]. Therefore enols are much more easily oxidized than the corresponding ketone tautomer. Supposing that the enolization is faster than the electron transfer, it ought to be possible to oxidize the enol present in small amounts beside the ketone in the equilibrium mixture. The following cyclization reactions are as useful approach to the chemistry of enol radical cations and can be considered as the α-umpolung of ketones.

Sterically hindered, mesityl-substituted, stable enols **72** have been examined with regard to one-electron oxidation. Using two equivalents of a one-electron oxidant such as triarylaminium salts, iron(III)phenanthroline, thianthrenium perchlorate or ceric ammonium nitrate in acetonitrile-benzofurans **73** are obtained in good yields within a few seconds [111].

Two possible mechanisms are proposed. Primarily the enol radical cation is formed. It either undergoes deprotonation because of its intrinsic acidity, producing an α-carbonyl radical, which is oxidized in a further one-electron oxidation step to an α-carbonyl cation. Cyclization leads to an intermediate cyclohexadienyl cation. On the other hand, cyclization of the enol radical cation can be faster than deprotonation, producing a distonic radical cation, which, after proton loss and second one-electron oxidation, leads to the same cyclohexadienyl cation intermediate as in the first reaction pathway. After a 1,2-methyl shift and further deprotonation, the benzofuran is obtained. Since the oxidation potentials of the enols are about 0.3–0.5 V higher than those of the corresponding α-carbonyl radicals, the author prefers the first reaction pathway via α-carbonyl cations [112]. Under the same reaction conditions, the oxidation of 2-mesityl-2-phenylethenol **74** does not lead to benzofuran but to oxazole **75** in yields of up to 85 %. The oxazole **75** is generated by nucleophilic attack of acetonitrile on the α-carbonyl cation or the proceeding enol radical cation.

A radical-cation-type cyclization of a series of isoprenoid polyene acetates has been described recently by Demuth. In the presence of an electron acceptor

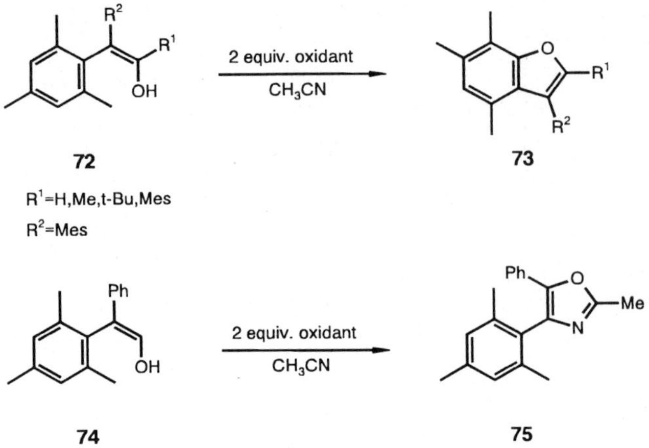

Scheme 16

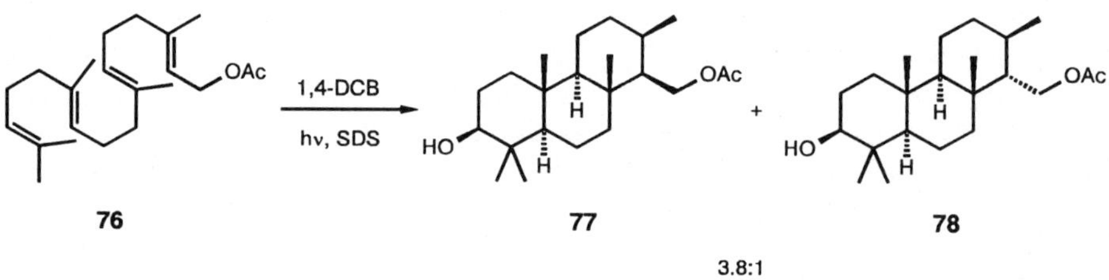

Scheme 17

in a heterogeneous medium, irradiation induces a consecutive bond-forming reaction cascade, building up arrays of stereogenic centers [113]. The author proposes a photochemically generated radical cation intermediate trapped by water in a stereoselective *anti*-Markovnikov sense. The acetate of all-*trans*-geranylgeraniol **76** in aqueous sodium dodecyl sulfate (SDS) solution in the presence of 1,4-dicyanobenzene (1,4-DCB) is transformed into the diastereomers **77** and **78** in the ratio of 3.8 : 1 with low yields (Scheme 17). Improved yields of up to 25% with an unchanged isomer ratio are achieved using the sensitizer couple 1,4-dicyanobenzene and phenanthrene. In anionic micellar medium the cyclization of homologues such as *trans*-geranyl acetate and all-*trans*-farnesyl acetate also takes place, whereas in an aqueous medium without SDS merely isomerization is observed. As a reason for the indispensibility of SDS an enhanced lifetime of the intermediate radical ion pair as well as a proper folding of the substrate is assumed.

2.2 Arene Radical Cations

One of the first examples reported of the oxidative intramolecular coupling reaction of alkoxy-substituted biarylalkanes via arene radical cations is the cyclization of 1,2-bis(3,4-dimethoxyphenyl)ethane **79**. It is converted into phenanthrene **80** by anodic oxidation [114–116] as well as with the chemical oxidant thallium(III)trifloroacetate (TTFA) in trifloroacetic acid (TFA) [117]. TTFA in TFA is shown to be a suitable oxidant for many biaryl systems, providing effectively a number of substituted tetrahydroisoquinolines and homoaporphine alkaloids. Treatment of electron-rich arylpropionic acids **81** with TTFA in TFA containing a small amount of BF_3·etherate results in the formation of dihydrocoumarins **82** and spirocyclohexadienone lactones **83** [118]. One-electron oxidation leads to the aromatic radical cation. The formation of spirocyclohexadienone lactones is presumed to occur via an intramolecular capture of this radical cation by the side-chain carboxyl group, further one-electron oxidation and final demethylation. For the formation of dihydrocoumarins two competitive pathways are conceivable. On the one hand, the radical cation can be trapped by the solvent, generating a radical. Intramolecular addition and rearomatization gives dihydrocoumarins. Alternatively, deprotonation of the radical cation followed by further one-electron oxidation can precede the cyclization.

Pandey and co-workers have generated arene radical cations by PET from electron-rich aromatic rings [119]. The photoreaction is apparently initiated by single-electron transfer from the excited state of the arene to ground state 1,4-dicyanonaphthalene (DCN) in an aerated aqueous solution of acetonitrile. Intramolecular reaction with nucleophiles leads to anellated products regiospecifically. The author explains the regiospecificity of the cyclization step from

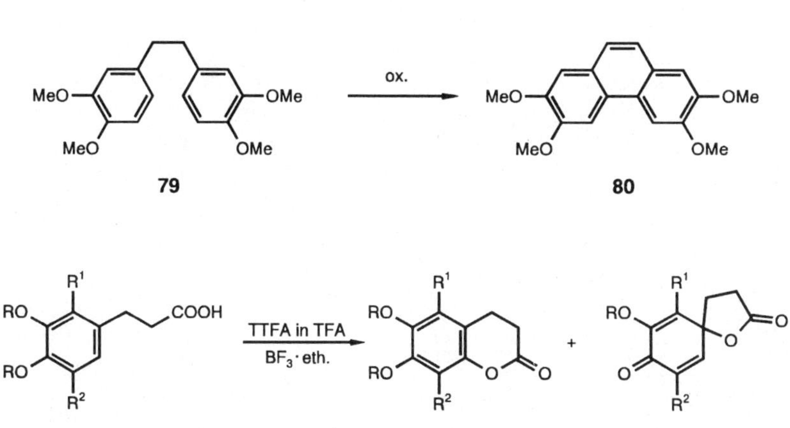

79 80 81 82 83

Scheme 18

FMO theory according to calculated electron densities at different carbon atoms of the HOMO of the radical cation (Hückel and MNDO programmes).

Using hydroxyl groups as nucleophiles, coumarins **86** are synthesized starting from the corresponding substituted cinnamic acids **84** in yields of 70 to 90% [120]. Cyclization of 2-aryl-1-alkyl-ethane-1-ol fails because of proton loss of the radical cation in the benzylic position followed by fragmentation. If the benzylic position is blocked by using the enolates **88** of 2-aryl-1-alkyl-ethane-1-ones **87**, 2-substituted benzofurans **89** are formed effectively [121]. Pandey's group has extended this method for the synthesis of Precocenes-I, a potent antijuvenile hormone compound [122]. Using amines as nucleophiles *N*-heterocycles can be built up efficiently as shown in Scheme 20. Cyclization of *β*-arylethylamines leads to highly substituted dihydroindoles in yields of up to 82%. This methodology has also been utilized to build up aromatic tricyclic *N*-heterocycles starting from **90**. A unique combination of two independent PET operation reactions by "wavelength switch" develops benzopyrrolizidine **92** similar to the mitomycin skeleton in a one-pot synthesis [123]. A recent publication of Pandey et al. illustrates the use of PET-generated arene radical cations for the intramolecular addition of enol silyl ethers **93** and **96** [124]. The attractive feature of this strategy lies in its ability to alter the ring size of the carboanellated product from a single starting compound (Scheme 21).

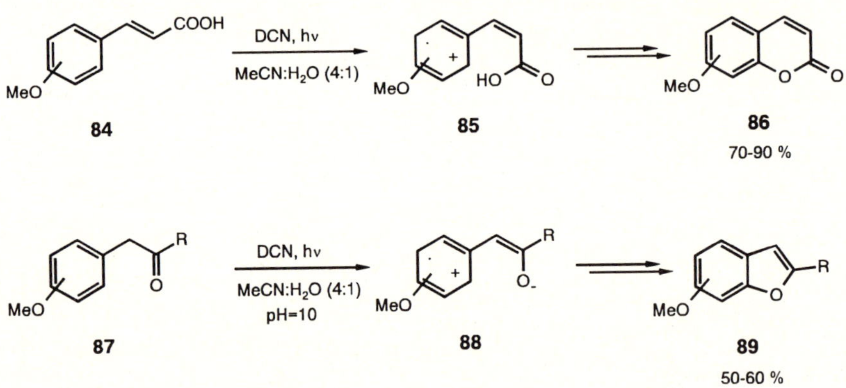

Scheme 19

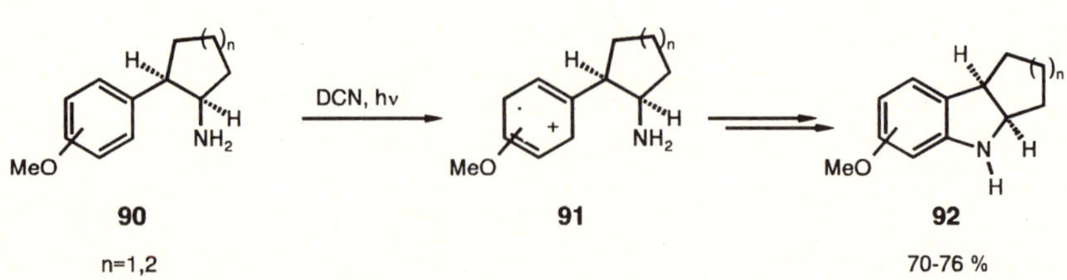

Scheme 20

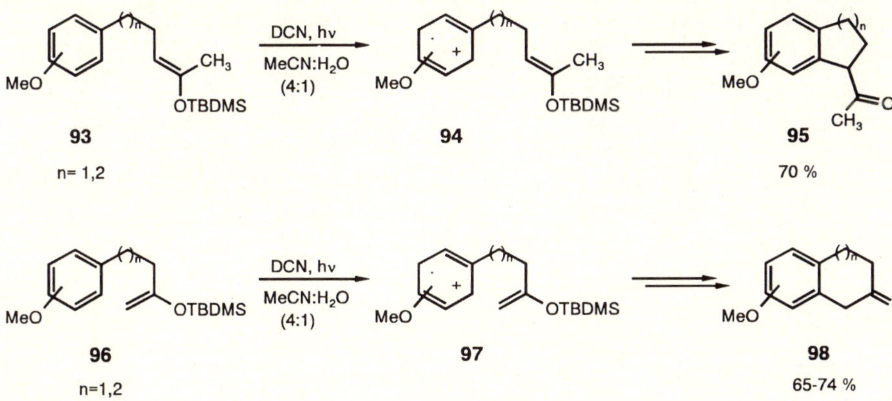

Scheme 21

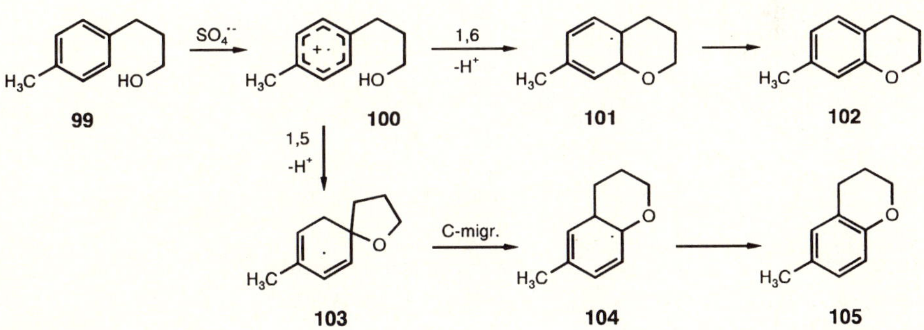

Scheme 22

McCleland has reported that 3-phenylpropan-1-ol [125] and 3-(*p*-methylphenyl)propan-1-ol **99** [126] cyclize to chromans when oxidized by the radical anion $SO_4^{-\bullet}$, generated by redox decomposition of $S_2O_8^{2-}$ with Fe^{2+}. The intermediate arene radical cation **100** is attacked by the nucleophilic hydroxy group. Whereas 1,6-cyclization yields 7-methylchroman **102**, 1,5-cyclization with subsequent C-migration leads to the regioisomer 6-methylchroman **105**. A dependence of the isomeric ratio and the combined yields to the pH value is determined. While 7-methylchroman **102** is the main product over a wide pH range, 6-methylchroman **105** is only formed at low pH. When the pH is lowered, the combined yields decrease due to the formation of an α-oxidized non-cyclized product.

A photochemical synthesis of isoquinoline and benzazepine derivatives in good preparative yields is shown in Scheme 23 [127, 128]. Upon electron-transfer-sensitized irradiation, the primary aminoethyl and aminopropyl stil-

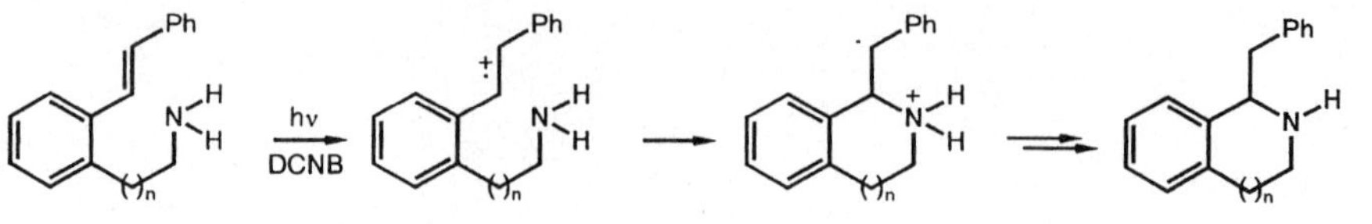

106
n=1,2

107

108

109

Scheme 23

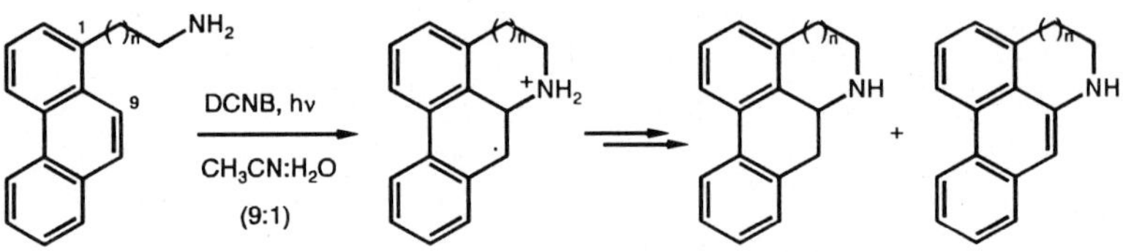

110	111	112	113
a n = 1		70 %	15 %
b n = 2		68 %	12 %

Scheme 24

bene **106** is converted into the stilbene radical cation **107**, followed by nucleophilic attack of the nitrogen. In both cases nucleophilic attack occurs regioselectively at the proximal end of the double bond, yielding comparably strain-free cyclization products **109**. Because of the high oxidation potential of the primary amine and a resulting endergonic electron transfer, direct irradiation of **106** does not lead to stilbene-amine-adducts but only to *cis/trans* isomerization. Cyclization of analogous secondary amines upon direct irradiation will be discussed in Sect. 4.1.

Lewis et al. have recently investigated the PET-oxidative cyclization reaction of several 1- and 9-(aminoalkyl)phenanthrenes [129] using *meta*-dicyanobenzene (DCNB) as sensitizer. Ring closure of the generated phenanthrene radical cation to the distonic radical cation is presumed to be the rate determining step in these reactions. In the case of 1-(aminoalkyl)phenanthrene **110**, nucleophilic addition of the nitrogen occurs to the C(9)–C(10)bond providing aporphine **112a** and hexahydrophenanthro[10,1-bc]azepine **112b**. The secondary product **113** is formed by photooxidation, since similar product ratios are obtained in the absence and presence of oxygen. Photolysis of 9-(aminoethyl)phenanthrene leads to a piperidine derivative according to a nucleophilic attack on the C(10), whereas by irradiation of 9-(aminopropyl)phenanthrene, the cyclization occurs at C(8) in preference to C(10).

2.3 Amine Radical Cations

Iminium cations serve as synthetically useful intermediates in the synthesis of various nitrogen heterocycles, especially those which possess diverse arrays of biological activity [130]. The photochemical generation of iminium cations from amines via amine radical cations and α-amino radicals represents an interesting application for a PET-sensitized reaction. Pandey et al. developed a sequential electron-proton-electron transfer route for in-situ generation of iminium cations by excited ^{1}DCN* in an aerated aqueous solution of acetonitrile [131, 132]. One-electron oxidation of amines leads to amine radical cation intermediates that can be deprotonated rapidly at the site adjacent to the nitrogen center. Further one-electron oxidation of the resulting α-amino radical produces the corresponding iminium cation, because the latter oxidation is expected to be an easy process due to the low ionization energy of the radical.

The iminium cation thus generated from unsymmetrical tertiary amines turns out to be highly regioselective, since the stereoelectronic factor subject to kinetic acidity influences the orientation of the deprotonation step of the initially formed amine radical cation. A direct approach to iminium cation chemistry is the intramolecular nucleophilic cyclization of *N*-substituted tertiary aminoalcohols **114** providing regio- and stereoselective tetrahydro-1,3-oxazines **117** and **118** in high yields [133, 134] (Scheme 25). The corresponding iminium cations **115** are trapped by the hydroxy group. It is presumed that the major diastereomer **117** arises from the propensity for a frontal attack of the hydroxy group due to steric reasons. As recently investigated by Cossy and coworkers, similar aminoalcohols can be cyclized photochemically using ketones, such as benzophenone or acetone, as electron acceptors instead of DCN [135].

An anodic azacyclization, producing tropane-related 11-substituted dibenzo[a,d]cycloheptimines **123**, was recently developed by Karady et al. [136, 137]. This two-electron process is initiated by anodic oxidation of the O-substituted hydroxylamine **119** in nucleophilic solvent. It is proposed that the first one-electron oxidation leads to the aminium radical cation **120** which adds rapidly to the double bond. The electron-rich carbon radical **121** is readily oxidized to the carbocation **122**. Selective nucleophilic attack on **122** from the less hindered *exo*-side yields the 11- substituted product **123**. Depending on the

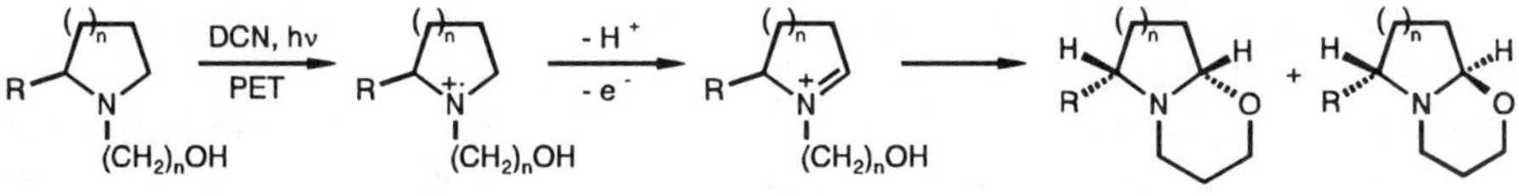

114 n=1,2 R=Me,n-Bu **115** **116** **117** **118** > 9:1

Scheme 25

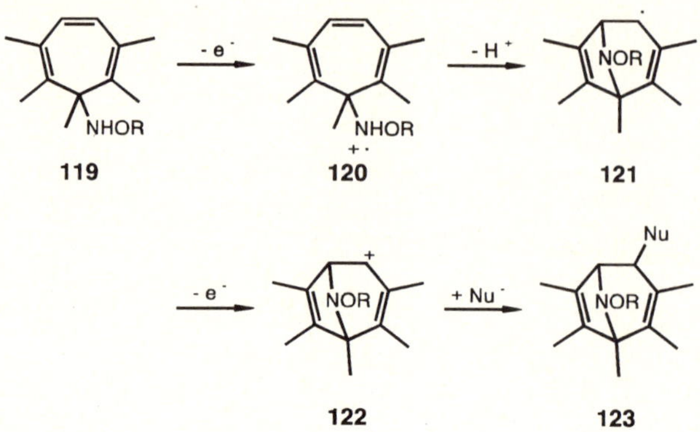

Scheme 26

solvent yields of 30 to 70% are obtained. A useful application of this azacyclization is the efficient synthesis of a hydroxylated metabolite of MK 801 [136, 137], which is itself an noncompetitive *N*-methyl-D-aspartate antagonist with an in-vivo antivulsant and neuroprotective activity [138, 139].

5-*exo*-Cyclizations of δ,ε-unsaturated aminium radical cation, formed by *N*-hydroxypyridine-2-thione (PTOC) carbamates under acidic conditions, are described by Newcomb et al. [58, 140, 141]. The aminium radical cation is shown to be significantly more efficient in cyclization than the neutral aminyl radical. According to the overall reaction sequence, the δ,ε-unsaturated aminyl radical **125**, generated from the PTOC carbamate precursor **124** in a radical chain reaction, is protonated in the presence of weak organic acid to give an aminium radical cation **126**. This aminium radical cation **126** cyclizes to give the carbon-centered radical **127**. When *tert*-butyl thiol is present in the reaction medium, it serves to trap **126** giving **129**, and in some cases, the thiol also reacts with the aminium radical cation **126** in competition with the cyclization to give amine **128**. In the absence of thiol, the carbon-centered radical reacts with the PTOC precursor to yield the 2-pyridyl alkyl sulfide **130**, the so-called self-trapped product. Cyclization of analogous *N*-chloramines and nitrosamines to pyrrolidines and piperidines, conducted under either strongly acidic or under mild acidic conditions, have been investigated earlier by Surzur and Stella [142–145].

A wide application of Newcomb's method provides a variety of *N*-heterocyclic systems, such as perhydroindoles, pyrrolizidines and aza-brigded bicycles [59, 60, 146]. The mild reaction conditions are compatible with several funtional groups of the substrate and several trapping agents to functionalize the cyclized product. 2-Substituted pyrrolizidines **132** are accessible by tandem cyclization of *N*-allyl-substituted PTOC carbamate **131**. In this case the allyl group on the nitrogen serves as an internal trap for the intermediate carbon radical. The *N*-methylcyclohept-4-enaminium radical cation, produced from the corres-

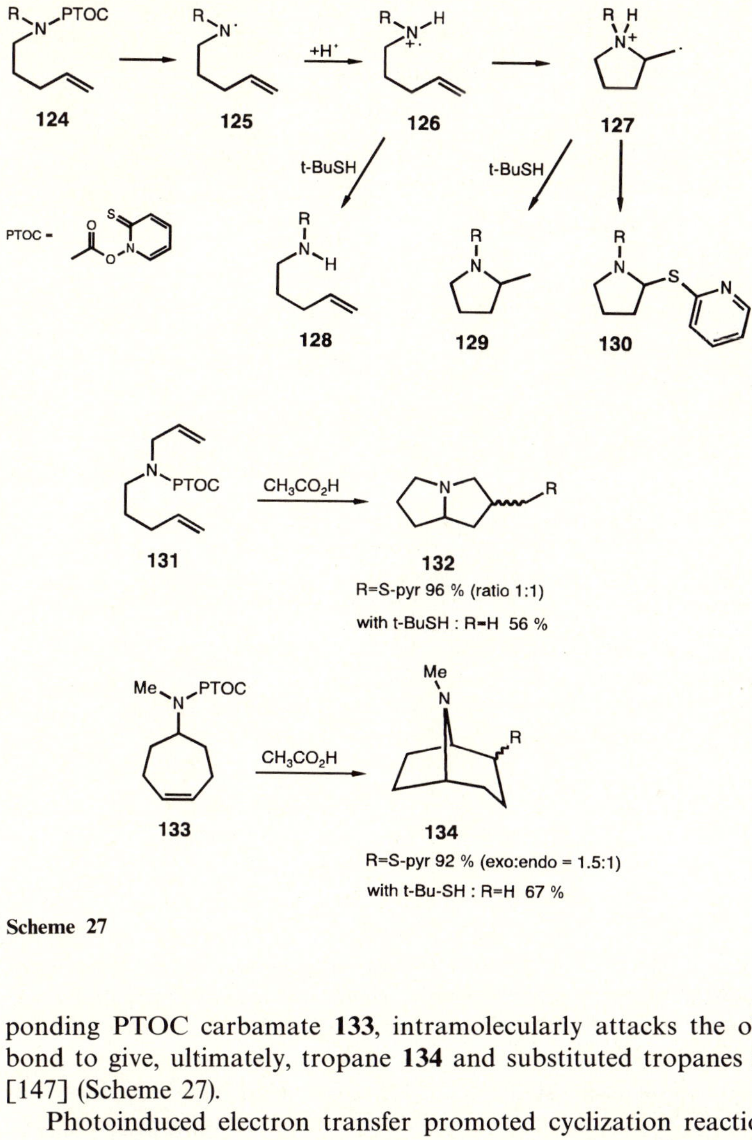

Scheme 27

ponding PTOC carbamate **133**, intramolecularly attacks the olefinic double bond to give, ultimately, tropane **134** and substituted tropanes in high yields [147] (Scheme 27).

Photoinduced electron transfer promoted cyclization reactions of α-silylmethyl amines have been described by two groups. The group of Pandey cyclized amines of type **135** obtaining pyrrolidines and piperidines **139** in high yields [148]. The cyclization of the α-silylated amine **140** leads to a 1 : 1 mixture of the isomers **141** and **142** [149]. The absence of diastereoselectivity in comparison to analogous 3-substituted-5-hexenyl radical carbocyclization stereochemistry [9] supports the notion that a reaction pathway via a free radical is unlikely in this photocyclization. The proposed mechanism involves delocalized α-silylmethyl amine radical cations as reactive intermediates. For stereochemical purposes, Pandey has investigated the cyclization reaction of **143**, yielding

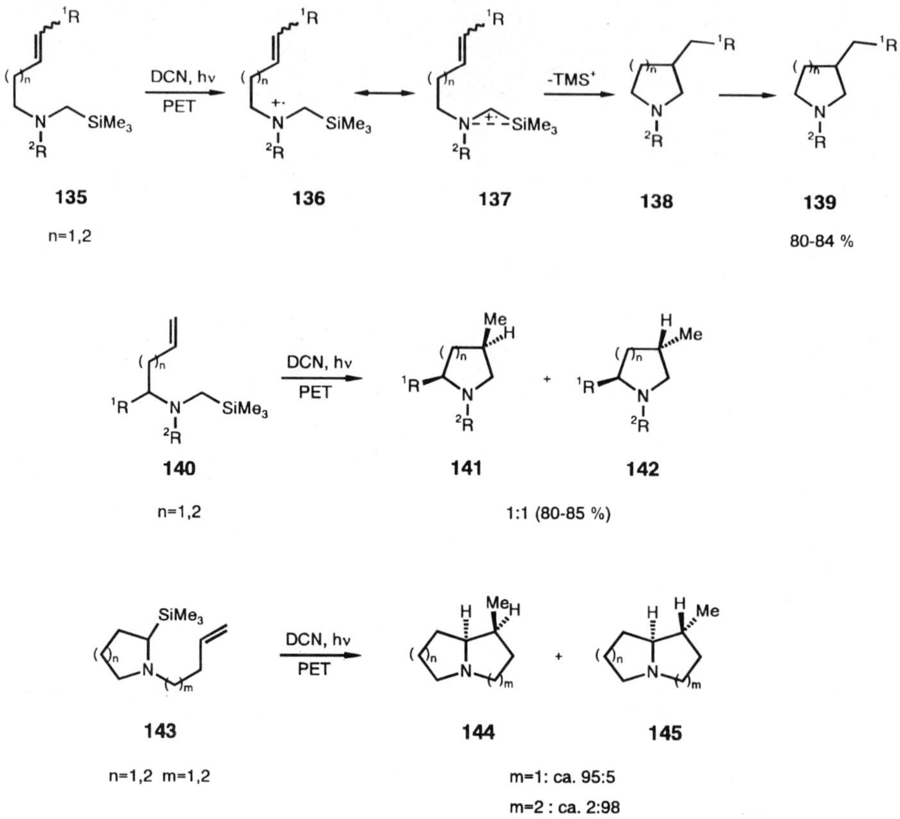

Scheme 28

1-azabicyclo[m +2.n +2.0] alkanes with high stereo-selectivity [132, 150]. When a 5-membered ring is formed, predominantly 1,5-*cis* stereochemistry is observed; whereas if 6-membered rings are produced, they show 1,6-*trans* stereochemistry.

Mariano's group have investigated the PET-generated intramolecular conjugate addition of α-silylmethyl amines linked to α,β-unsaturated ketones, such as enones and phenanthrenyl ketones. It represents an interesting application of PET-sensitized reactions for the synthesis of hydroisoquinoline derivatives in a regio- and stereoselective manner. The photocyclization of **146** merely leads to the *cis*- and *trans*-fused isomers **148** and **149**, whereas, as a single diastereomer, the tetracyclic amino ketone **152** is produced by cyclization of **150** [151]. Systems in which α-methylsilyl amines are linked to cyclohexadienones **153** undergo photocyclization reactions as well, producing *cis*-fused functionalized hydroisoquinolines only **154** [152]. In contrast to Pandey's assumption, Mariano proposes a reaction pathway via α-amino radicals. He assumes that

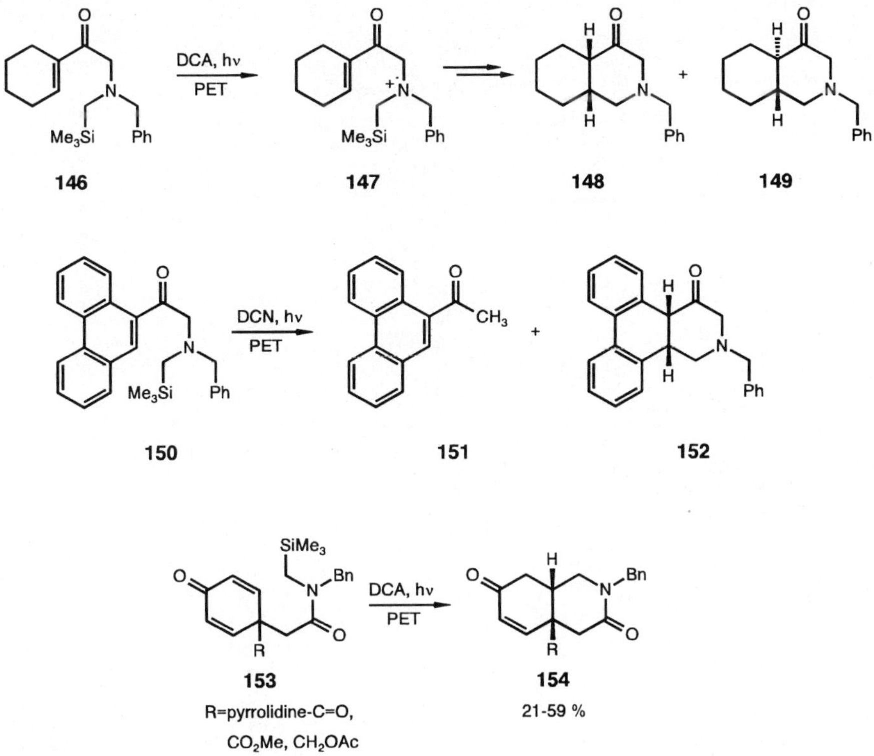

Scheme 29

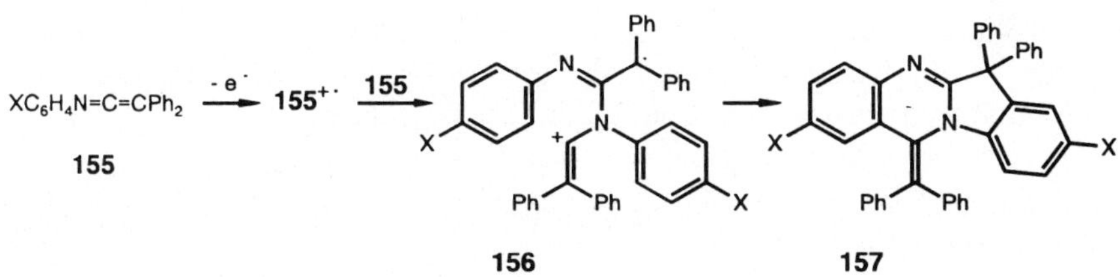

Scheme 30

desilylation of the PET-generated amino radical cation takes place before the cyclization step. Direct irradiation of amine-enone pairs investigated by Mariano will be discussed in Sect. 4.3.

2.4 Ketene Imine Radical Cations

Though the chemical oxidation of aryl-substituted ketene imines with various oxidizing agents [153–156] generally leads to cleavage products, anodic oxida-

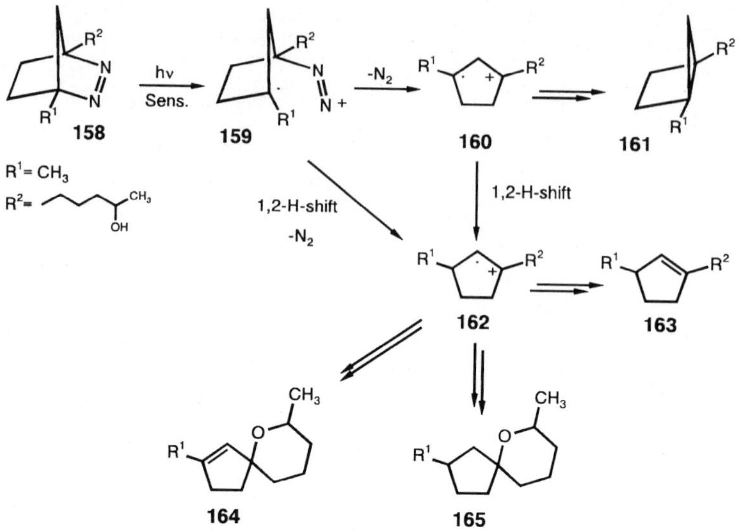

Scheme 31

tion offers the possibility of cyclic dimerization between a ketene imine radical cation and its parent molecule. Complex heterocyclic compounds are produced [157, 158]. A nucleophilic attack of the ketene imine **155** to a ketene imine radical cation gives the distonic radical cation **156** (only one conceivable intermediate is shown). The further reaction pathway involves two attacks on the adjacent aromatic rings, either one electrophilic and one radical, or two electrophilic after a second one-electron oxidation (Scheme 30).

2.5 Diazenyl Radical Cations

Although cyclic azoalkanes are well known as biradical precursors [159] they have been used as 1,2- and 1,3-radical cation precursors only recently [160–164]. Apart from the rearrangement products bicyclopentane **161** and cyclopentene **163**, the PET-oxidation of bicyclic azoalkane **158** yields mostly unsaturated spirocyclic products [165]. Common sensitizers are triphenylpyrylium tetrafluoroborate and 9,10-dicyanoanthracene with biphenyl as a cosensitizer. The ethers **164** and **165** represent trapping products of the proposed 1,2-radical cation **162**. Comparison of the PET chemistry of the azoalkane **158** and the corresponding bicyclopentane **161** additionally supports the notion that the non-rearranged diazenyl radical cation **159** is involved (Scheme 31).

3 Radical Anion Cyclizations

The previous chapter covered radical cation cyclization reactions that were a consequence of single-electron oxidation. In the following section, radical anion cyclization reactions arising from single-electron reduction will be discussed. In contrast to the well documented cyclization reactions via carbon-centered free radicals [3, 4], the use of radical anions has received limited attention. There are only a few examples in the literature of intramolecular reductive cyclization reactions via radical anions other than ketyl. Photochemically, electrochemically or chemically generated ketyl radical anions tethered to a multiple bond at a suitable distance, have been recognized as a promising entry for the formation of carbon-carbon bonds.

3.1 Ketyl Radical Anions

The electroreductive cyclization reaction of 6-heptene-2-one **166**, producing *cis*-1,2-dimethylcyclopentanol **169**, was discovered more than twenty years ago [166]. In agreement with Baldwin's rules, the 5-*exo* product is obtained in a good yield. Since that time, the mechanism of this remarkable regio- and stereoselective reaction has been elucidated by Kariv-Miller et al. [167-169]. Reversible cyclization of the initially formed ketyl radical anion **167** provides either the *cis* or the *trans* distonic radical anion. Subsequent electron transfer and protonation from the kinetically preferred **168** leads to the major *cis* product **169**. The thermodynamically preferred **170** is considered as a source of the trace amounts of the *trans* by-product **171** (Scheme 32).

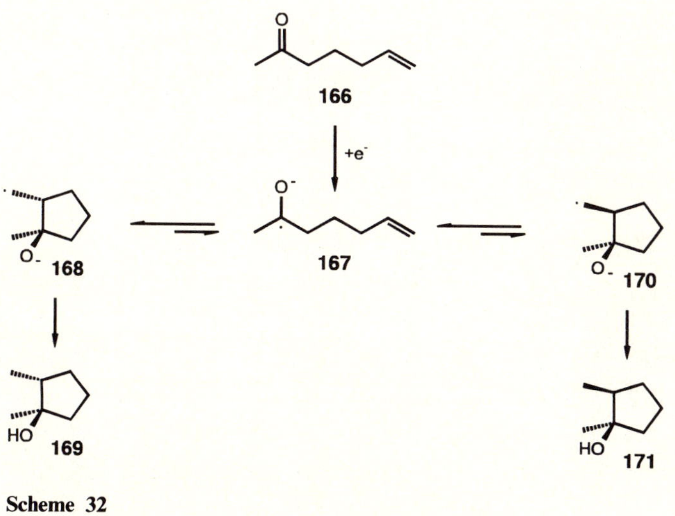

Scheme 32

Intensive studies concerning the photoreductive cyclization of distinct ketones and aldehydes are made by Cossy et al. [170]. They describe how bicyclic tertiary cycloalkanols **173** and **174** can be prepared from δ,ε-unsaturated ketones **172** in good yields, initiated by photoinduced electron transfer from triethylamine in acetonitrile or by photoionization in pure hexamethylphosphoric triamide (HMPA) [171, 172]. The reaction is stereo-, chemo- and

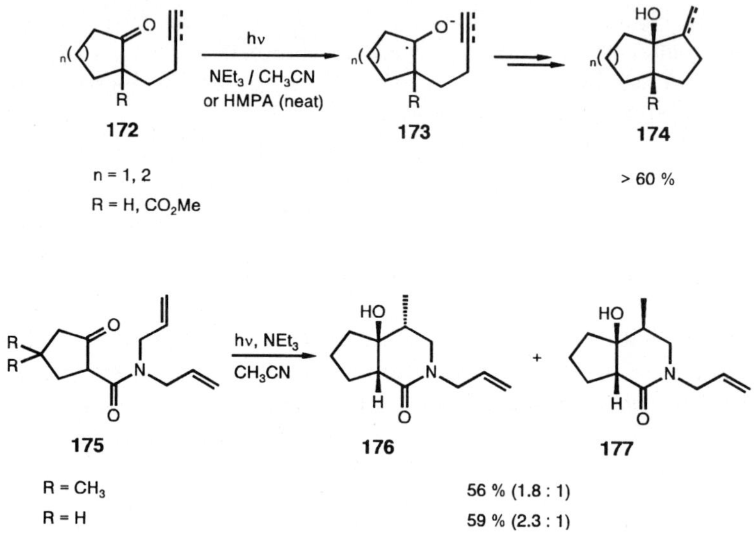

Scheme 33

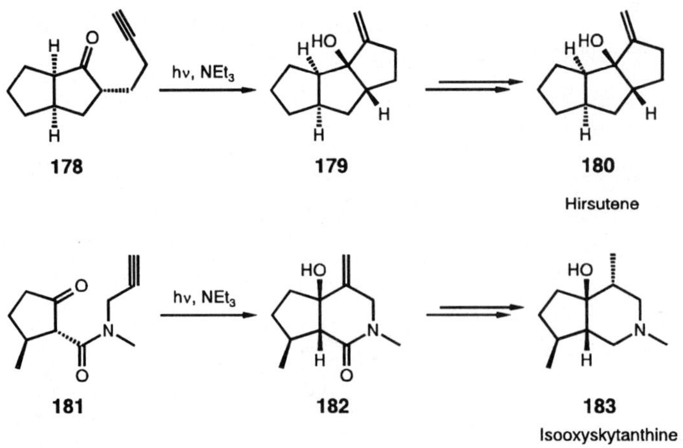

Scheme 34

regioselective. The *exo*-trig or *exo*-dig cyclized products are obtained exclusively. This methodology has also been successfully applied to the synthesis of *N*-heterocycles [173–175], such as δ-lactams **176** and **177**, starting from *N,N*-diallyl-β-oxoamides **175**. As shown earlier by Shono et al. δ,ε- and ε,ζ-unsaturated ketones similar to **172** also cyclize under electroreductive conditions with remarkable regio- and stereoselectivity. In a mixed solvent of methanol and dioxane or in *N,N*-dimethylformamide using carbon electrodes, *cis* isomers of exocyclic tertiary alcohols **174** are exclusively obtained in yields of 32 to 87% [176, 177]. Due to the stereoselectivty and the mild conditions, the PET-reductive cyclization reaction has been applied to the total synthesis of polycyclic and heterocyclic biologically active natural products such as (±) Hirsutene **180** [178, 179] or (±) Isooxyskyanthine **183** [180]. Further interesting applications have been reported by Cossy et al. [181–183].

Apart from PET-reductive cyclization, chemical reduction has also been applied to the total synthesis of natural products such as capnellenediol **186** [184]. Naphthalene sodium is shown to be a suitable oxidant for generating ketyl radical anions which cyclize efficiently in a 5-*exo*-dig mode. In contrast, electroreductive cyclization of **184** does not lead to **185**, but exclusively to the thermodynamically preferred 5-*exo* isomer with a remaining double bond in the endocyclic position [185] (Scheme 35). The steroid precursor 4.5-secocholestan-5-one **187**, in which the 10α-side chain is varied, has been cyclized under the same conditions [186–188] (Scheme 36). Reduction with naphthalene sodium or sodium in ether exclusively produces the A:B-*cis* steroid **188** with an *exo* double

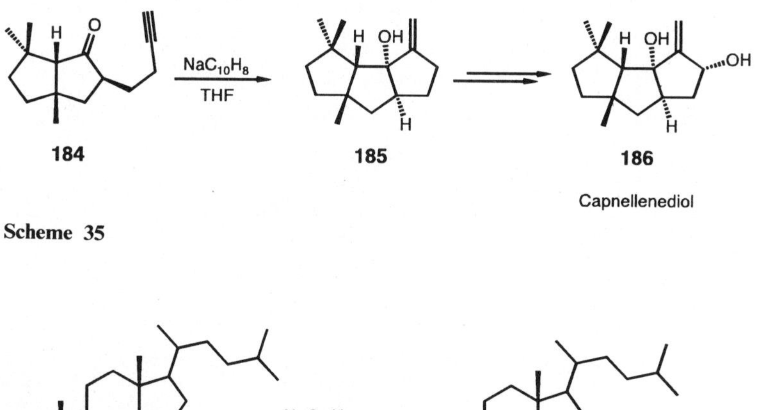

Scheme 35

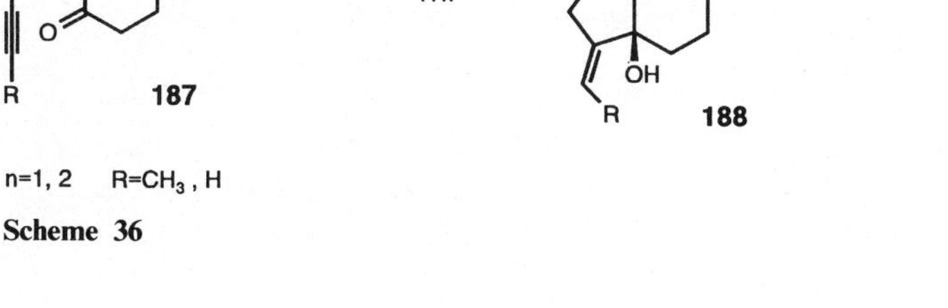

n=1, 2 R=CH_3, H

Scheme 36

bond; whereas, if alkali metal in ammonia is used, the cyclization is accompanied by simple reduction to a secondary alcohol. In the first step, the mechanism certainly involves one-electron reduction to form the ketyl radical anion. Simple reduction of the ketone with sodium in ammonia is explained by the powerful reducing agent in which two-electron reduction, leading to the vicinal dianion, is about as fast as the cyclization step [70].

A series of bicyclo[3.3.0]octanols are accessible by electroreductive tandem cyclization of linear allyl pentenyl ketones **189**, as shown by Kariv-Miller et al. [189]. The electrolyses are carried out with an Hg-pool cathode and a Pt-flag anode. As electrolyte, tetrabutylammonium tetrafluororborate is used. The reaction is stereoselective, yielding only two isomers **192** and **193**. In a competing reaction, a small amount of the monocyclic alcohol is formed. Since all the monocycles have the 1-allyl and the 2-methyl group in *trans* geometry it is assumed that this terminates the reaction. The formation of a bicyclic product requires that the first cyclization provides the *cis* radical anion which leads to *cis*-ring juncture [190] (Scheme 37).

An interesting application of photoinduced electron transfer involving fragmentation and subsequent cyclization reactions is reported by Kirschberg and

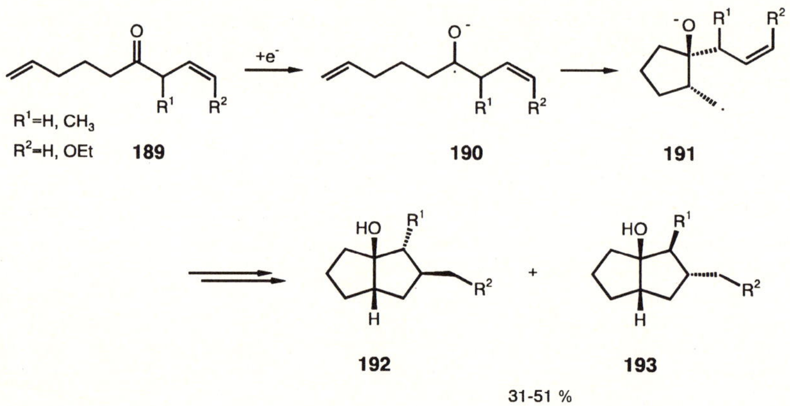

Scheme 37

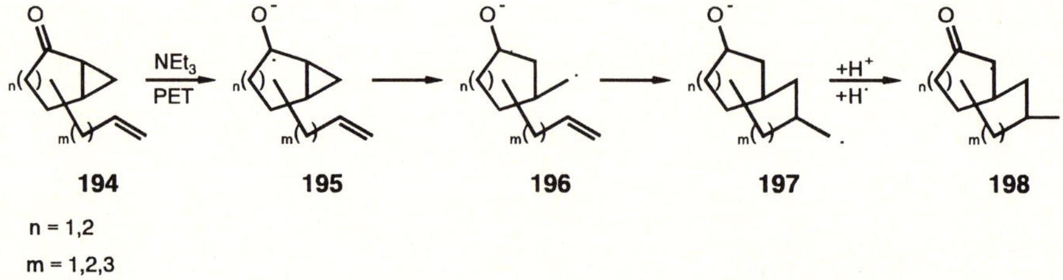

Scheme 38

Mattay [191]. Irradiation of unsaturated bicyclo[4.1.0]heptanones and bicyclo-[3.1.0]hexanones **194** in the presence of triethylamine in acetonitrile ($\lambda = 300$ nm) leads to ketyl radical anions **195** which are subjected to a regioselective cleavage of the cyclopropane moiety. The distonic radical anion intermediate **196** containing a side chain double bond cyclizes efficiently. By this methodology various types of ring anellated and spirocyclic compounds are accessible (Scheme 38). Further studies on the regiochemistry and mechanism of ketyl radical anion intramolecular cyclization reactions have been by Newmark and co-workers. Cathodic reduction of (*Z*,*E*)-4,8-cyclododecadien-1-one **199** at a constant current, catalyzed by *N*,*N*-dimethylpyrrolidinium tetrafluoroborate leads to bicyclic alcohols **202** and **204** in an 5-*exo*-trig mode [192]. As evidence for the rapid cyclization of **200**, only a minute amount of non-cyclized alcohol is formed. At different stages of conversion, various product ratios are observed, indicating that the cyclization of **200** to **201** and **203** is reversible. The final product distribution reflects combined kinetic and thermodynamic control with some preference for the former, since in any case a mixture of **202** and **204** predominantly containing **202** is obtained.

The electroreductive cyclization of ketones and aldehydes linked to α,β-unsaturated esters **205** has been investigated by Little and co-workers. Good

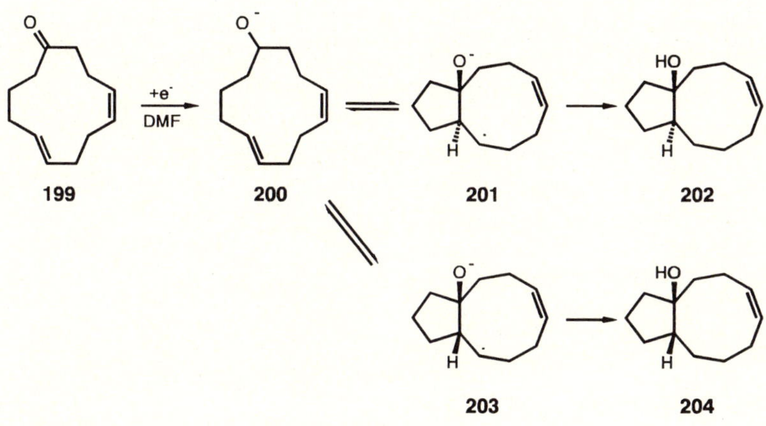

Scheme 39

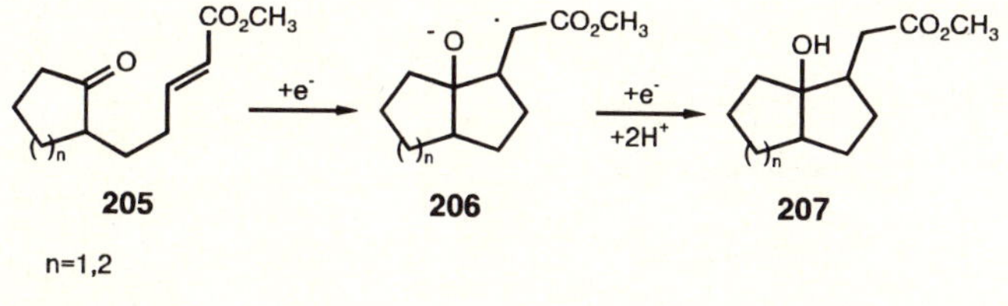

Scheme 40

yields of mono- and bicyclic γ-hydroxy esters **207** are obtained [193, 194]. According to the proposed mechanism, the radical anion formed by initial electron transfer undergoes a rapid reversible cyclization to the closed form of the radical anion **206**. Then by an irreversible proton transfer to oxygen and a second electron transfer, the enolate is generated followed by rapid protonation. To elucidate the mechanism, intensive studies using linear sweep voltammetry have been made [195]. Similar starting materials for reductive cyclization mediated by magnesium in dry methanol have been used by Lee et al. [196]. The authors also suggest the involvement of radical anions.

Cyclization of terminal allenic ketones is shown to be highly regio- and stereoselective using either electroreductive conditions or chemical reduction with naphthalene sodium [185, 197]. The allenic ketone **208** cyclizes efficiently in an 5-*exo*-trig mode, producing vinyl-substituted bicyclooctanol **211** in yields of 37% under electroreductive conditions and 23% after chemical reduction. Electroreductive cyclization of ketone **213**, with the length of the allenic side chain shortened, leads to kinetically stable *endo*-ene bicyclic alcohol **212**, whereas chemical reduction produces thermodynamically stable *exo*-ene bicyclic alcohol **214**. Similar compounds have successfully been cyclized under PET-reductive conditions, yielding 5-*exo* products containing the double bond in *endo*- and exocyclic positions with some preference for the former [172] (Scheme 41).

Shono and Kise have investigated the electroreductive coupling reaction of γ- and δ-cyanoketones, yielding bicyclic α-hydroxy ketones **218** and their dehydroxylated equivalents **221** [198]. Optimized yields are obtained when the electroreduction is carried out in *i*-propanol at a controlled potential of -2.8 V using a divided cell equipped with a ceramic diaphragm and an Sn or Ag cathode. The product ratio is controlled by the reaction temperature. When the reaction is carried out at 25°C, almost exclusively the α-hydroxy ketone **218** is obtained, whereas at 65°C the obviously thermally dehydroxylated ketone **221** is the predominant product (Scheme 42). Furthermore, this methodology has been

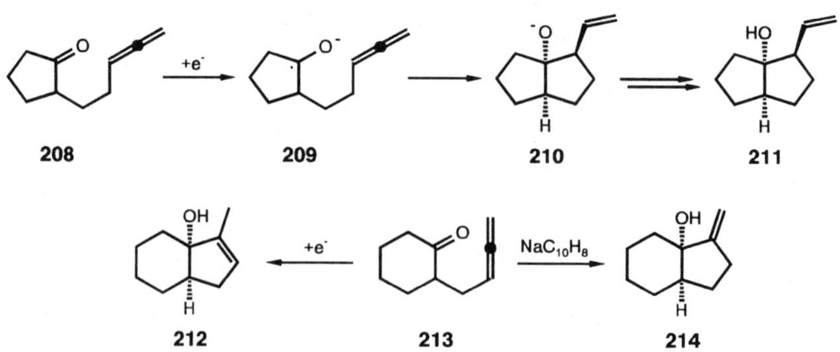

Scheme 41

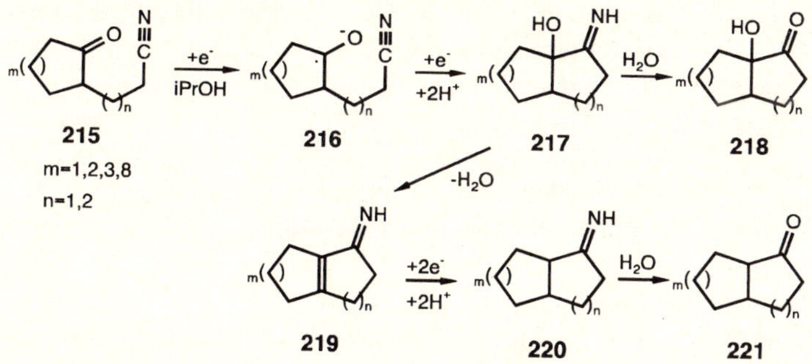

Scheme 42

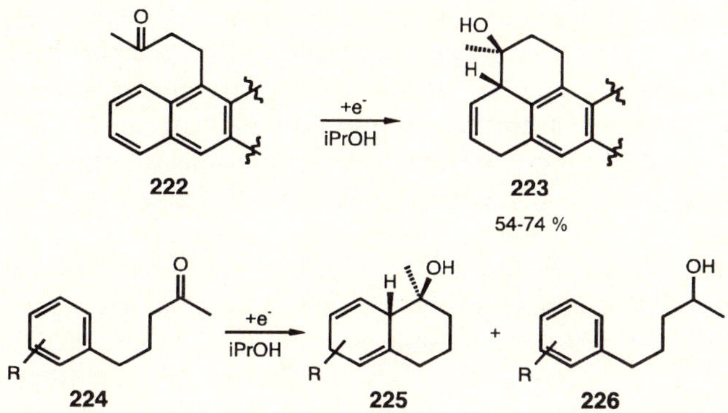

Scheme 43

applied to the synthesis of several natural products, such as guaiazulene, valeranone and rosaprostol [199].

A variety of bi- and polycyclic compounds have been synthesized by electroreductively induced intramolecular coupling of β- and γ-aryl ketones **222** and **224** [200, 201]. The authors discuss a mechanism involving a ketyl radical anion which attacks the aromatic ring to give the *cis* rather than the *trans* intermediate, due to the electronic repulsion between the anionic oxygen atom and π-electrons of the phenyl group. A control-experiment with butylbenzene under the same conditions indicates that another reaction pathway via arene radical anions is unlikely. A remarkable difference in the product ratio is observed when different electrolytes are used. Whereas no cyclized product is formed when lithiumperchlorate is used, in the presence of tetraalkylammonium salts, Et_4NOTs or Bu_4NClO_4 **225** is obtained in yields of up to 70%. The authors explain these results according to the covalent or ionic nature of the bond between the ketyl radical anion and the counter ion, respectively. In the case of

Li$^+$, the covalent character predominates and further reduction is faster than ring closure. The covalent character decreases when quaternary ammonium cations are used as counter ions. Hence, the subsequent electron transfer is much slower than the cyclization step and only the cyclized product is obtained.

3.2 Radical Anions of *α,β*-Unsaturated Carbonyl Compounds

Little has investigated monoactivated and doubly activated alkenes tethered to butenolide with respect to electroreductive cyclization [202]. The geminally activated systems **227** undergo cyclization to diastereomeric products **228** and **229** in an 1:1 mixture, whereas both the α,*β*-unsaturated monoester and α,*β*-unsaturated mononitrile fail to cyclize. Only saturation of the C–C double bond of butenolide is observed. The author explains these results by distinct reactivity and lifetime of the intermediate radical anions. The radical anions derived from the monoactivated olefins are less delocalized than those of **227** and therefore should be shorter lived and more reactive. In this case preferential saturation occurs. The radical anions derived from the doubly activated alkene **227** are comparatively long-lived and less basic and thus capable of attacking the C–C double bond of the butenolide moiety. A decrease in saturation, accompanied by a marked increase in cyclization, is observed.

The eletrochemical reductive cyclization of α,*β*-unsaturated esters **230** and **232** bearing a mesylate leaving group have been investigated by Gassman et al. [203, 204]. Although *trans*-configurated **232** undergoes cyclization in a yield of 60% the *cis*-configurated diastereomer fails to cyclize. Therefore the authors have suggested that the α,*β*-unsaturated ester **232** is reduced to a radical anion which performs a classical backside S_N2 displacement on the mesylate anion. With *cis* stereochemistry, this type of displacement cannot occur.

Recently it has been shown that radical anionic cyclization of olefinic enones effectively compete with intramolecular [2 + 2]-cycloaddition to form spirocyclic compounds [205, 206]. 3-Alkenyloxy- and 3-alkenyl-2-cyclohexenones **235** are irradiated in the presence of triethylamine. As depicted in Scheme 46 two reaction pathways may operate. Both involve electron transfer steps, either to the starting material (resulting in a direct cyclization) or to the preformed cyclobutane derivative **239**, which undergoes reductive cleavage. The second

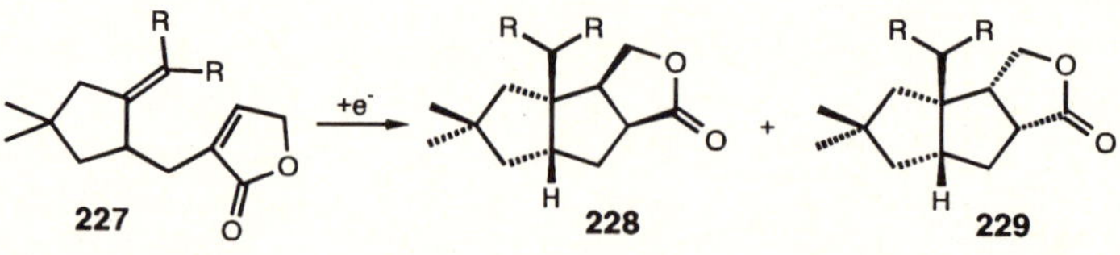

Scheme 44

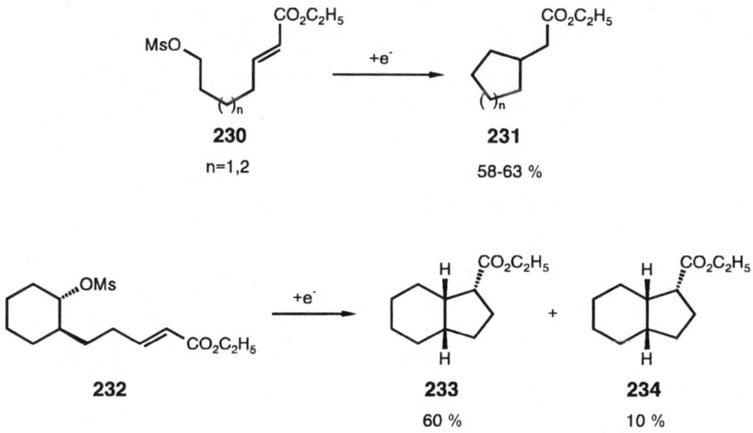

Scheme 45

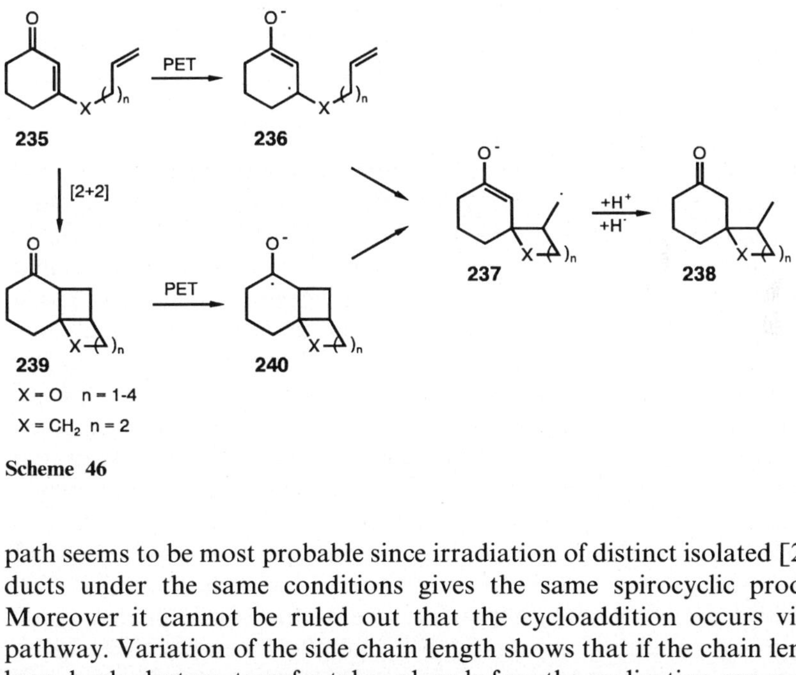

Scheme 46

path seems to be most probable since irradiation of distinct isolated [2 + 2]-adducts under the same conditions gives the same spirocyclic products **238**. Moreover it cannot be ruled out that the cycloaddition occurs via a PET-pathway. Variation of the side chain length shows that if the chain length is too long, back-electron transfer takes place before the cyclization can occur. Short chains enable fast cycloaddition, producing only spirocyclic products. A new photosystem for the generation of enone radical anions has been designed by Pandey and Harjra [207]. A solution of 9,10-dicyanoanthracene (DCA), triphenylphosphine (Ph_3P) and enone **241** in *N*,*N*-dimethylformamide is irradiated ($\lambda = 405$ nm). Based on different control-experiments and calculations upon the free energy ΔG_{ET} of the involved electron transfers a reaction pathway

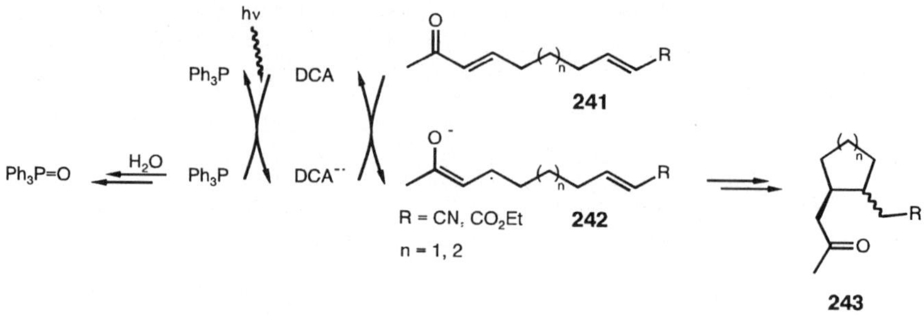

Scheme 47

utilizing DCA as an electron relay is proposed. As shown in Scheme 47 electron transfer from Ph_3P to excited DCA leads to $Ph_3P^{+\cdot}$ and $DCA^{-\cdot}$. Consequent back-electron transfer from $DCA^{-\cdot}$ to **241** yields the radical anion **242**, which undergoes cyclization to form **243** in good yields. DCA is regenerated.

3.3 Other Radical Anions

A cyclization reaction of the quinonemethide precursor **244** linked to a 5-hexenyl group is described by Dimmel [208]. The quinonemethide radical anion **246** is considered to be a as reactive intermediate. In 1 M NaOH at 135°C in the presence of five equivalents of glucose or two equivalents of anthrahydroquinone, **245** is converted into the quinonemethide radical anion **246**. Consequent cyclization to a five-membered ring (**249**) occurs. If glucose is used as an additive, even a second cyclization is observed, yielding three unique tricyclo[7.3.0.0]dodecatrienes **250–252** (Scheme 48). Apparently glucose reduces the distonic radical anion intermediate relatively slowly, providing time for cyclization to take place. Generally, this reaction gives additional information for elucidating the nature of chemical reactions that occur during the pulping of wood, such as the anthrahydroquinone-induced lignin fragmentation.

Amatore et al. recently discovered that electrochemical oxidation of metallated phenyl prenyl sulphone **253** leads to an unexpected cyclic dimer **258** [209, 210]. A radical-anion coupling mechanism is proposed. Since in this anodic oxidation process, the radical anion **255** cannot be generated from **254** in a medium containing no reducing species (divided cell), it should arise from a coupling reaction between a phenyl prenly sulfone anion and its corresponding radical. The feasibility of this reaction on a preparative scale has been demonstrated by the almost quantitative transformation of **254** into **258** by the slow addition of catalytic amounts of the reducing agent sodium anthracene. Cyclic voltametry experiments additionally support the presence of the radical anion **255** as a key intermediate.

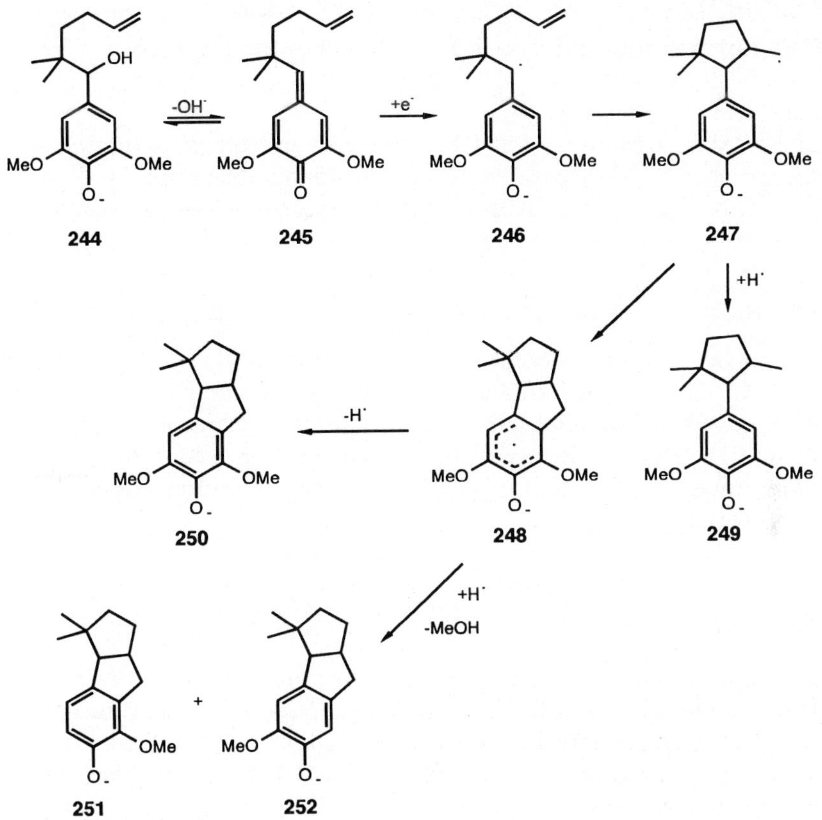

Scheme 48

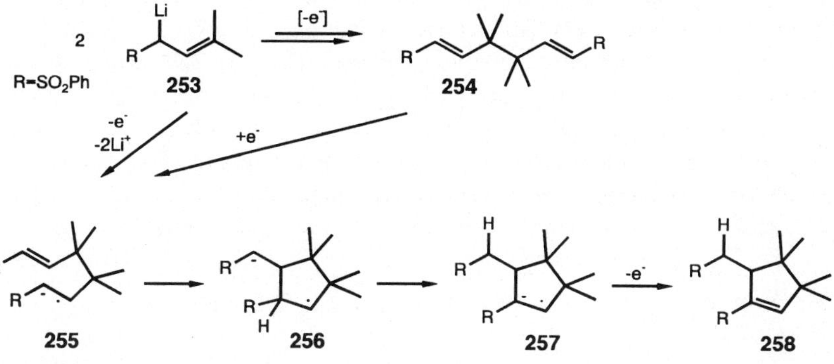

Scheme 49

4 Cyclization Reactions Involving Radical Cations and Radical Anions in Linked Donor-Acceptor Systems

In an electron donor-acceptor system (D-A) in which an electron donor (D) and an electron acceptor (A) are separated by a carbon chain, the exciplex state may be considered as a resonance hybrid of the electron transfer configuration ($D^{+\bullet}A^{-\bullet}$) mixed with the locally excited configuration (D*A) or (DA*) [211].

$$\text{D-(CH}_2)_n\text{-A} \xrightarrow{h\nu} \{\text{D}^{+\bullet}\text{-(CH}_2)_n\text{-A}^{-\bullet}\}^* \longrightarrow \overset{\frown}{\text{D-(CH}_2)_n\text{-A}} \qquad (5)$$

The photolysis of donor-acceptor systems shows a reaction pattern of unique synthetic value. Direct irradiation of the donor-acceptor pairs, such as arene-amine, leads by intramolecular electron transfer, to amine radical cations and arene radical anions. The generated radical cation and radical anion intermediates undergo cyclization reactions providing efficient synthetic routes to *N*-heterocycles with a variety of ring sizes.

4.1 Amine-Arylalkene

Intramolecular coupling reactions between the acceptor-donor pair styrene-amine have been intensively studied by Lewis et al. [212–215]. High conversions are possible due to the absorption of Pyrex-filtered light ($\lambda > 300$ nm) by the styrene chromophore, but not by the styrene-amine adducts. According to the proposed mechanism, direct irradiation of ω-(β-styryl)-α-(methylamino)alkanes **259** leads to the locally excited styrene singlet state which is quenched by electron transfer from the ground state amine to singlet styrene. N-H transfer to either C-α or C-β of the styrene double bond generates two different biradical intermediates, **261** and **263**, which combine. The regioselectivity of the N-H transfer depends on the polymethylene chain length. Predominantly the less-strained ring sizes are built up. As shown in Scheme 51, direct irradiation of secondary aminoethyl and aminopropyl stilbene **265** and **269** leads to benzazepines **268** and **272** in improved yields, compared to previous synthetic routes to these molecules [127, 128]. In both cases, regiospecific hydrogen transfer forms 1,7-biradical intermediates. Lewis et al. presume that the regiospecifity arises from the geometrical constraints placed upon the intermediate exiplexes by the alkyl chain connecting the stilbene and the amine.

4.2 Amine-Arene

Further investigations on the acceptor-donor system arene-amine have been made by Sugimoto et al. They have reported that direct irradiation of 9-(ω-anilinoalkyl)phenanthrenes give spirocyclic pyrroline derivatives, invoked by

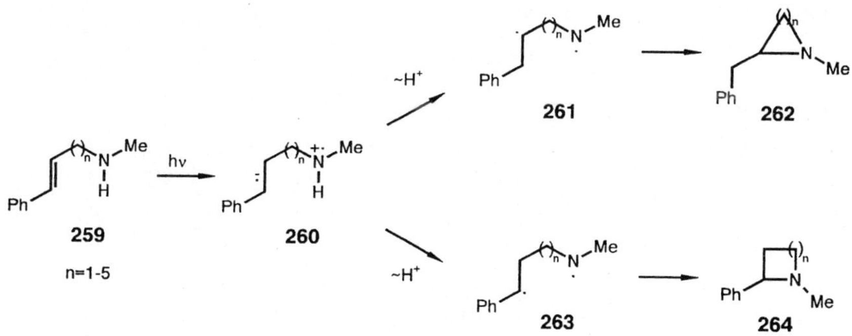

Scheme 50

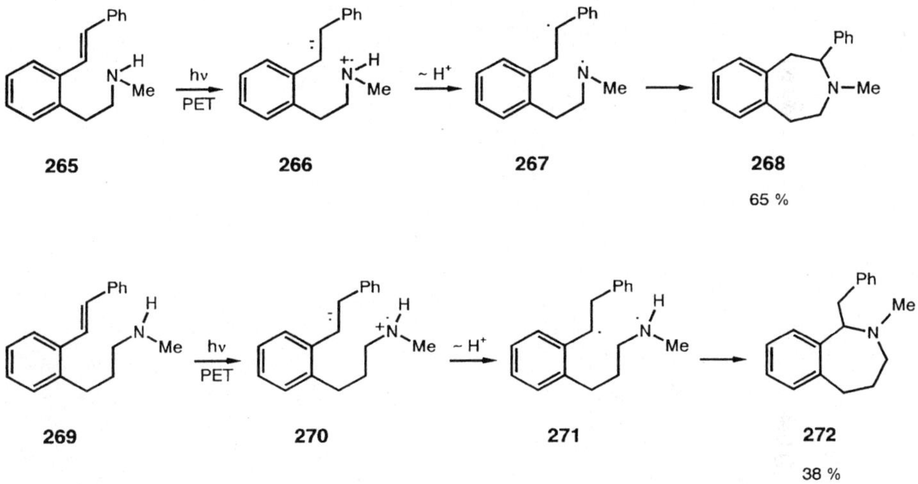

Scheme 51

N-H addition to the phenanthrene C(9)–C(10) bond [216–218]. Recently, photolysis of 9-(6-anilinohexyl)phenanthrene **273** in benzene yielded an unexpected polycyclic product **274** which has a benzomorphan skeleton [219]. The reaction is explained as proceeding by cyclization of the anilino group to the 6-position of the phenanthrene ring. After hydrogen migration and consecutive photoaddition of the N-H group to the C(7)–C(8) double bond, **274** is formed. The authors could not discern whether the initial reaction step is the electron transfer from the anilino chromophore to the phenanthrene moiety, since, in a non-polar solvent such as benzene, exciplexes formed by interaction between the excited phenanthrene and the anilino group are also conceivable. Fluorescence-quenching experiments with various 9-(aminomethyl)phenanthrenes have shown that in this case electron transfer does take place [216].

As depicted in Scheme 53 the photo-Smiles rearrangement involves radical ion pairs. Intermediately the spiro-type Meisenheimer complex **277** is formed

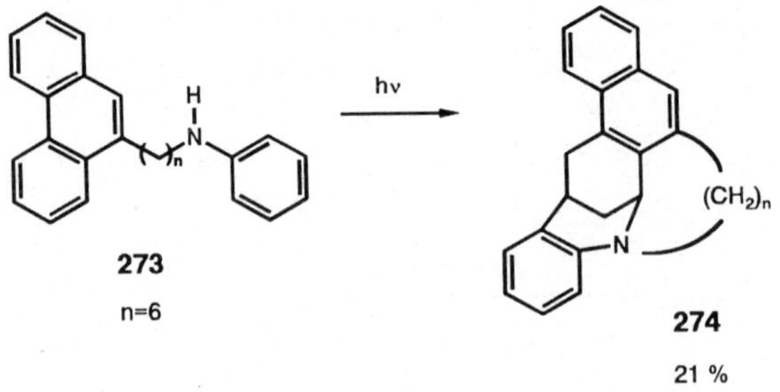

Scheme 52

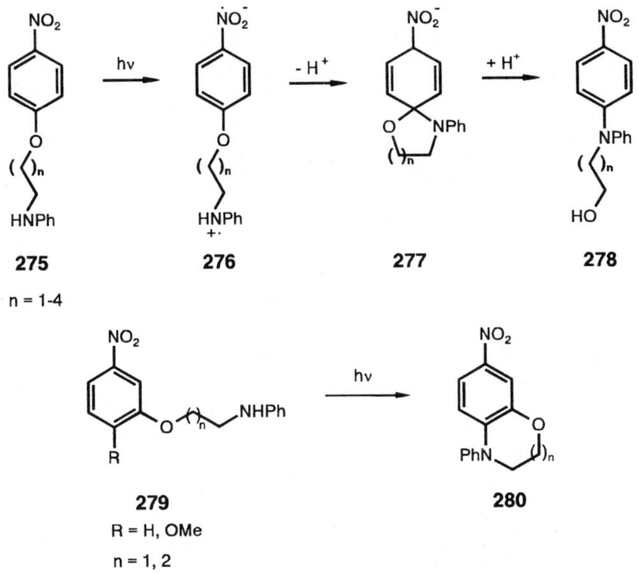

Scheme 53

[220, 221]. The mechanism has been ascertained by laser flash photolysis. Besides aromatic substitution, it is possible to obtain cyclic products by using meta-substituted compounds **279** due to the para-directing effect of the nitro group [222, 223].

4.3 Amine-Enone

The direct photocyclization of another interesting acceptor-donor pair, the amine-enone system, has been reported by Mariano [224–226]. Direct irradiation of β-(aminoethyl)cyclohexenones **281** leads to the excitation of the conjugated cyclohexenone chromophore. Intramolecular single-electron transfer from the amine donor to the cyclohexenone excited state results in the formation of the zwitterionic biradical **282**. α-Deprotonation and combination of the intermediate biradical **283** leads to spiro *N*-heterobicyclic systems **284**. Mariano has investigated in detail the effects of substituents R_1 and R_2 on the kinetic acidity of the amine radical cation, which is reflected in the distribution of the products obtained [227]. If *N*-(trimethylsilyl)methyl substituents are involved, desilylation as well as deprotonation of the intermediate zwitterionic biradical is conceivable. In this case, the solvent appears to govern the chemoselectivity. Specifically, desilylation is preferred in the polar protic solvent methanol, whereas deprotonation is favored in aprotic acetonitrile.

Further research on intramolecular photocyclization of amino enones and amino ketones based on electron transfer has been made by Kraus and Chen [228]. In analogy to the earlier results of Roth and El Raie [229], Kraus and Chen obtain the cyclopropanol derivative **286** as single stereoisomer by direct irradiation of **285**. Photolysis of the amino enone **287** does not lead to a three-membered ring product, but only to pyrrolidine **288**. The irradiation of the unsaturated keto ester **289** results in the even more unexpected formation of a nine-membered ring product **290**. Such remote photocyclizations have rarely been described so far.

4.4 Amine-Ketone

Besides Kubo's investigations, Hasegawa et al. have attained photochemical-promoted medium-sized ring systems from (dialkylamino)ethyl β- and γ-oxoesters **291**. The reaction is presumed to occur via electron transfer from the nitrogen atom to the excited carbonyl group and subsequent remote hydrogen transfer. In none of the photoreactions of (dialkylamino)ethyl β-oxoesters [230, 231] are the ε-hydrogen transfer products, aminolactones, obtained. Compared to ε-hydrogen abstraction via a stereochemically unfavorable eight-membered transition state, η-hydrogen abstraction via a less strained ten-membered cyclic transition state is preferred. The photocyclization of (dialkylamino)ethyl γ-oxoesters [232] also does not produce aminolactones, but rather exclusively nine-membered azalactones. Nevertheless γ-oxoesters show a different photochemical behaviour due to their smaller conformational flexibility. The photo-

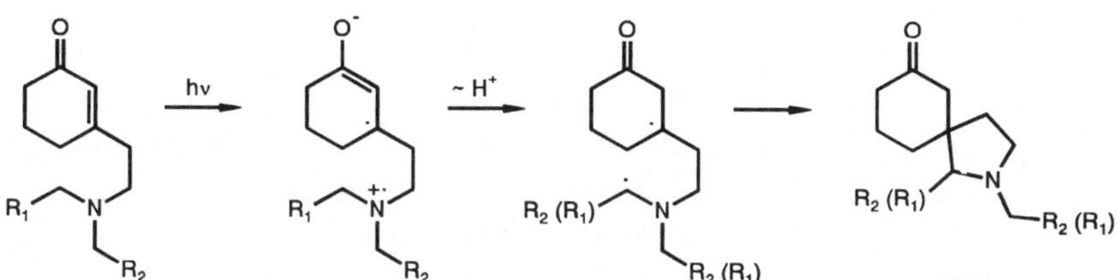

281 **282** **283** **284**

Scheme 54

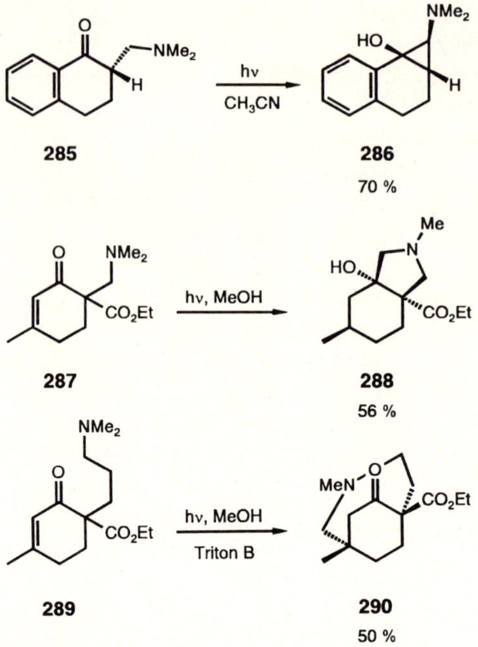

Scheme 55

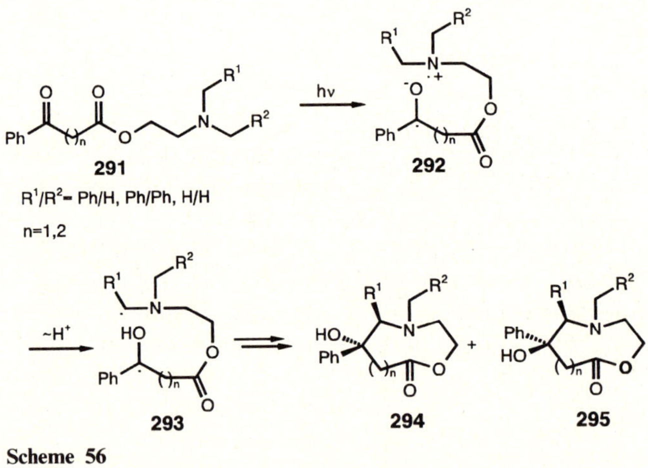

Scheme 56

cyclization requires a biradical with a stable alkyl radical center such as a benzyl radical. The cyclization of γ-oxoester containing one *N*-benzyl and one *N*-methyl group only leads to azalactones through benzylic proton migration. Unstable biradicals derived from (dimethylamino)ethyl γ-oxoester are subject to polymerization rather than to a change in conformation that makes them suitable for cyclization.

4.5 Alkene-Imide

The photochemistry of imides, especially of the *N*-substituted phthalimides, has been studied intensively by several research groups during the last two decades [233–235]. It has been shown that the determining step in inter- and intramolecular photoreactions of phthalimides with various electron donors is the electron transfer process. In terms of a rapid proton transfer from the intermediate radical cation to the phthalimide moieties the photocyclization can also be rationalized via a charge transfer complex in the excited state.

According to Kubo's investigations, solvent-incorporated cyclization of a wide variety of *N*-(2-alkenyl)- and *N*-(3-alkenyl)-phthalimides [236–238] and *N*-alkenylphthalimides with a more remote alkenyl double bond [239] provide new five-, six- and medium-sized ring systems. Irradiation in methanol-acetonitrile triggers off intramolecular electron transfer from the olefin double bond to the excited phthalimide carbonyl group. Due to the high nucleophilic character of the methanol, the olefin radical cation can be trapped. The invoked radical combines with the radical anion to yield coupled products. Kubo has extended the methanol-incorporated cyclization to arenedicarboximides with varying arene structure, such as naphthalene and phenanthrene derivatives **296** [240], showing that this photocyclization technique has general synthetic utility. The examination of the fluorescence spectra, *cis/trans* isomerization and the free-energy change assoziated with the electron transfer support the suggested reaction pathway.

4.6 Amine-Imide

The same process has been observed by irradiating phthalimides that are *N*-substituted with an alkyl amine [241, 242]. In particular, the photochemical cyclization of ω-anilinoalkylphthalimides **301** reveals great synthetic potential. It has been successfully applied to a wide range of alkyl chain lengths, producing medium-sized and large ring systems in yields of 6 to 20% [243, 244] (Scheme 58). Moreover, the C–C bond formation is exclusively observed between the imide carbonyl and the *N*-methyl group. Such regioselective remote photocyclization shows that hydrogen transfer is not only possible from the γ-position of the imide carbonyl via a six-membered transition state, but also, at least formally, from remote positions. As an exception, in the photolysis of ω-anilinobutylphthalimides no regioselectivity is attained. The cyclic products **302** and **303** are formed in yields of 6 and 5%, respectively, after hydrogen transfer from the two carbon atoms adjacent to the nitrogen. This observation gives rise to the assumption that 1,7-and 1,9-proton transfer are about equally probable.

4.7 Cyclopropane-Imide

An example of an intermolecular photocyclization is described by Mazzocchi [245, 246]. *N*-methylphthalimide **304** is irradiated in the presence of phenylcyc-

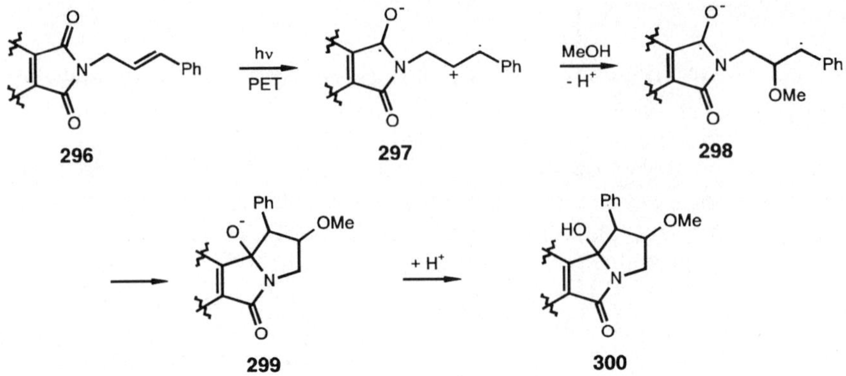

Scheme 57

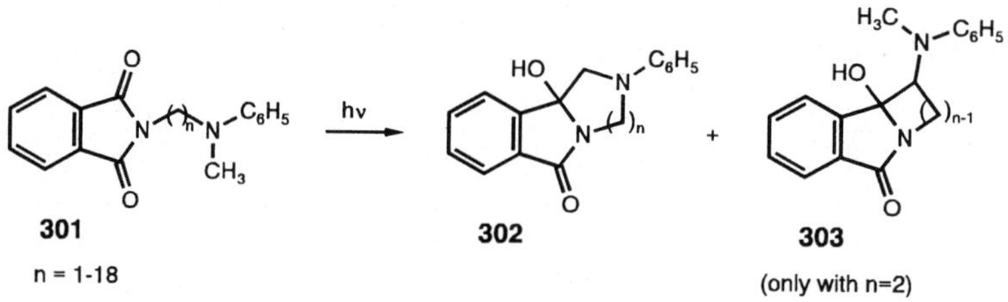

Scheme 58

lopropane **305** in acetonitrile. The mechanism of the reaction occurs via electron transfer from phenylcyclopropane to the imide, followed by coupling of the radical ion pair at the 1,2-position of the carbonyl to the cyclopropane ring. Deuterium-labeling studies support the assumption that the cycloaddition takes place stepwise rather than concerted [247]. Two isomeric spirocyclic products are formed in a 1:1 mixture in yields of 11%, respectively. Improved yields (57% combined) with the same isomeric ratio are obtained using *N*-methylnaphthalimide as an electron acceptor. In methanol as solvent, the cycloadduct **309** is not obtained, but instead a methanol-incorporated non-cyclized adduct is formed due to the low nucleophilic character of the imide radical anion compared to methanol itself.

4.8 Thioether-Imide

As shown by Sato et al. *N*-phthaloyl derivatives of *C*-unprotected amino acids efficiently undergo decarboxylation upon irradiation [248, 249]. In this case, the *N*-phthaloyl α-amino acid of methionine **310** represents an exception, because the normal decarboxylation route is not followed. Two main products are obtained, the *trans*-hydroxy acid **313** and the tetracyclic lactone **314** [250, 251]

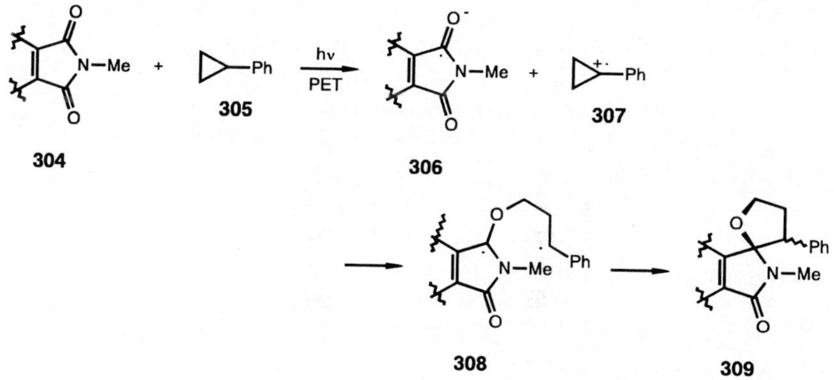

Scheme 59

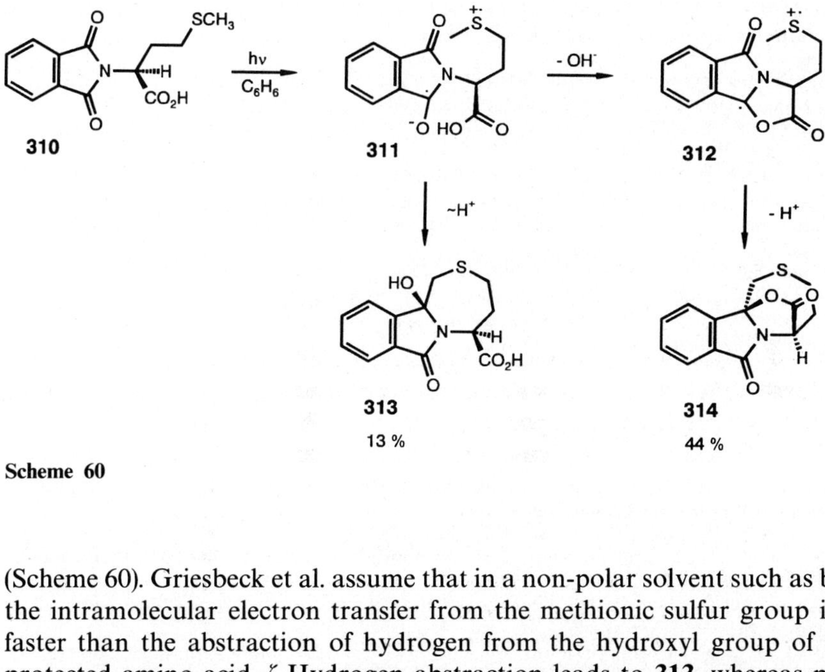

Scheme 60

(Scheme 60). Griesbeck et al. assume that in a non-polar solvent such as benzene the intramolecular electron transfer from the methionic sulfur group is much faster than the abstraction of hydrogen from the hydroxyl group of the unprotected amino acid. ζ-Hydrogen abstraction leads to **313**, whereas previous lactonization of the zwitterionic biradical **311** yields **314**. Since the *cis*-hydroxy acid is not detected it is conceivable that it cyclizes immediately to the lactone **314**. Photolysis of the corresponding methyl ester under the same conditions attains improved yields (84% combined) of two diastereomeric tricyclic products in a ratio of 48:52.

Futhermore, the photocyclization of the donor-acceptor pair thioether-imide has already been applied to the synthesis of the berberine alkaloid chilinene as an α-key step [252].

5 References

1. Thebtaranonth C, Thebtaranonth Y (1990) Tetrahedron 46: 1385
2. Thebtaranonth C, Thebtaranonth Y (1994) New Directions in Organic and Biological Chemistry: Cyclization Reactions. CRC Press, London
3. Giese B (1986) Radicals in Organic Synthesis: Formation of Carbon-Carbon Bonds. Pergamon Press, Oxford
4. Curran DP (1988) Synthesis 417 and 489
5. Beckwith ALJ (1993) Chem Soc Rev 22: 143
6. Julia M (1971) Acc Chem Res 4: 386
7. Porter NA, Magnin DR (1986) J Am Chem Soc 108: 2787
8. Baldwin JE (1976) J Chem Soc, Chem Commun 734
9. Beckwith ALJ (1981) Tetrahedron 37: 3073
10. RajanBabu TV (1991) Acc Chem Res 24: 139
11. Roth HD (1992) Top Curr Chem 163: 131
12. Saeva FD (1990) Top Curr Chem 156: 59
13. Maslak P (1993) Top Curr Chem 168: 1
14. Bauld NL, Belville DJ, Pabon R, Chelsky R, Green G (1983) J Chem Soc 105: 2378
15. Bauld N, Belville DJ, Hairirchian B, Lorenz KT, Pabon RA, Reynolds DW, Wirth DD, Chiou H-S, Marsh BK (1987) Acc Chem Res 20: 371
16. Mattay J (1989) Synthesis 233
17. Mattay J (1987) Angew Chem Int Ed Engl 26: 825
18. Fukuzumi S, Kochi JK (1982) Tetrahedron 38: 1035
19. Mattes SL, Farid S (1982) Acc Chem Res 15: 80
20. Ledwith A (1972) Acc Chem Res 5: 133
21. Neunteufel RA, Arnold DR (1973) J Am Chem Soc 95: 4080
22. Shigemitsu Y, Arnold DR (1975) J Chem Soc, Chem Commun 407
23. Maroulis AJ, Shigemitsu Y, Arnold DR (1978) J Am Chem Soc 100: 535
24. Maroulis AJ, Arnold DR (1979) Synthesis 819
25. Maroulis AJ, Arnold DR (1979) J Chem Soc, Chem Commun 351
26. Arnold DR, Maroulis AJ (1977) J Am Chem Soc 99: 7355
27. Bunnett JF, Creary X (1975) J Org Chem 40: 3740
28. Bunnett JF (1978) Acc Chem Res 11: 413
29. Kornblum N (1975) Angew Chem Int Ed Engl 14: 734
30. Rossi RA, de Rossi RH (1983) Aromatic Substitution by the $S_{RN}1$ Mechanism, ACS Monograph 178, Washington DC
31. Savéant J-M (1990) Adv Phys Org Chem 26: 1
32. Kornblum N (1982) in Patai S (ed) The Chemistry of Amino, Nitroso and Nitro Compounds and their Derivatives, part 1. Wiley, New York, p 361
33. Bowman WR (1988) Chem Soc Rev 17: 283
34. Bowman WR, Lackson SW (1990) Tetrahedron 46: 7313
35. Eberson L (1987) Electron Transfer Reactions in Organic Chemistry. Springer, Berlin Heidelberg NewYork
36. Kochi JK (1991) in Trost BM, Fleming I (eds) Comprehensive Organic Synthesis, vol 7. Pergamon Press, Oxford, p 849
37. Mijs WJ, de Jonge CRHI (eds) (1986) Organic Synthesis by Oxidation with Metal Compounds. Plenum Press, New York
38. Wiberg KB (ed) (1965) Oxidation in Organic Chemistry. Academic Press, New York
39. Trahanovsky S (ed) (1973) Oxidation in Organic Chemistry. Academic Press, New York
40. Trahanovsky S (ed) (1982) Oxidation in Organic Chemistry. Academic Press, New York
41. Arndt D (1981) Manganese (VII, VI, V, IV, III) Compounds as oxidizing Agents in Organic Chemistry. Open Court Publishing, LaSalle IL
42. Sheldon RA, Kochi JK (1981) Metal-catalyzed Oxidations of Organic Compounds. Mechanistic Principles and Synthetic Methodology Including Biochemical Processes. Academic Press, New York
43. Lund H, Baizer MM (eds) (1991) Organic Electrochemistry. Marcel Dekker, New York
44. Torii S (1985) Electroorganic Synthesis: Methods and Applications: Part I – Oxidations. VCH, Deerfield Beach FL

45. Ross SD, Finkelstein M, Rudd EJ (1975) Anodic Oxidation, Academic Press, New York
46. Yoshida K (1984) Electrooxidation in Organic Chemistry: The Role of Cation Radicals as Synthetic Intermediates. Wiley, New York
47. Schäfer HJ (1981) Angew Chem Int Ed Engl 20: 911
48. Schäfer HJ (1990) Top Curr Chem 152: 91
49. Shono T (1984) Electroorganic Chemistry as a New Tool in Organic Synthesis. Springer, Berlin Heidelberg New York
50. Kavarnos GJ, Turro NJ (1986) Chem Rev 86: 401
51. Fox MA, Chanon M (eds) (1988) Photoinduced Electron Transfer, Part A-D. Elsevier, Amsterdam
52. Mattay J (ed) (1990-1994) Top Curr Chem, vol 156,158,159,163,168,169
53. Mariano PS (ed) (1991 and 1992) Advances in Electron Transfer Chemistry, vols 1–3. JAI Press, Greenwich Connecticut
54. Shida T, Haselbach E, Bally T (1984) Acc Chem Res 17: 180
55. Lehman TA, Bursey MM (1976) Ion Cyclotron Resonance Spectrometry
56. Turner DW, Baker AD, Baker C, Brundle CR (1970) Molecular Photoelectron Spectroscopy. Wiley, NewYork
57. Chow YL (1980) in Abramovitch RA (ed) Reactive Intermediates, vol 1. Plenum Press, New York, p 151
58. Newcomb M, Deeb TM, Marquardt DJ (1990) Tetrahedron 46: 2317
59. Newcomb M, Marquardt DJ, Deeb TM (1990) Tetrahedron 46: 2329
60. Newcomb M, Marquardt DJ, Kumar, MU (1990) Tetrahedron 46: 2345
61. McClelland BJ (1964) Chem Rev 64: 301
62. Huffman JW (1983) Acc Chem Res 16: 399
63. Maslak P, Narvaez JN, Kula J, Malinski DS (1990) J Org Chem 55: 4550
64. Saveant JM (1980) Acc Chem Res 13: 323
65. Kariv-Miller E, Pacut RI, Lehman GK (1988) Top Curr Chem 148: 97
66. Rossi RA (1982) Acc Chem Res 15: 164
67. Holy NL (1974) Chem Rev 74: 243
68. Caine D (1976) Organic Reactions 23:1
69. Rautenstrauch V, Geoffrey M (1977) J Am Chem Soc 99: 6280
70. Pradhan SK (1986) Tetrahedron 42: 6351
71. Ashby EC, Deshpande AK, Patil GS (1995) J Org Chem 60: 663
72. Grimshaw J, Haslett RJ, Trocha-Grimshaw J (1977) J Chem Soc, Perkin Trans 1 2448
73. Koppang MD, Ross GA, Woolsey NF, Bartak DE (1986) J Am Chem Soc 108: 1441
74. Guo QX, Qin XZ, Wang JT, Williams F (1988) J Am Chem Soc 110: 1974
75. Shono T, Nishiguchi I, Kashimura S, Okawa M (1978) Bull Chem Soc Jpn 51: 2181
76. Ito Y, Konoike T, Saegusa T (1977) J Am Chem Soc 99: 1487
77. Kobayashi Y, Taguchi T, Morikawa T, Tokuno E, Sekiguchi S (1980) Chem Pharm Bull 28: 262
78. Baciocchi E, Casu A, Ruzzioni R (1989) Tetrahedron Lett 30: 3707
79. Moriarty RM, Penmasta R, Prakash I (1987) Tetrahedron Lett 28: 873
80. Ito Y, Konoike T, Saegusa T (1975) J Am Chem Soc 97: 649
81. Fujii T, Hirao T, Ohshiro Y (1992) Tetrahedron Lett 33: 5823
82. Moriarty RM, Prakash I, Duncan MP (1985) J Chem Soc, Chem Commun 420
83. Snider BB, Kwon T (1990) J Org Chem 55: 4786
84. Snider BB, Kwon T (1992) J Org Chem 57: 2399
85. Heidbreder A, Mattay J (1992) Tetrahedron Lett 33: 1973
86. Curran DP, Chang CT (1989) J Org Chem 54: 3140
87. Hintz S (1994) Diploma Thesis, Münster University, Germany
88. Hintz S, Mattay J, unpublished results
89. Heidbreder A (1994), Ph.D. Thesis, Münster University, Germany
90. Julia M, Maumy M (1969) Bull Soc Chim Fr 2415 and 2427
91. Moeller KD, Marzabadi MR, New DG, Chiang MY, Keith S (1990) J Am Chem Soc 112: 6123
92. Moeller KD, Tinao LV (1992) J Am Chem Soc 114: 1033
93. Hudson CM, Marzabadi MR, Moeller KD, New DG (1991) J Am Chem Soc 113: 7372
94. Moeller KD, Hudson CM (1991) Tetrahedron Lett 32: 2307
95. Moeller KD, Hudson CM, Tinao-Wooldridge LV (1993) J Org Chem 58: 3478
96. Hudson CM, Moeller KD (1994) J Am Chem Soc 116: 3347
97. Moeller KD, New DG (1994) Tetrahedron Lett 35: 2857

98. Cossy J, Bouzide A, Leblanc C (1993) Synlett 202
99. Cossy J, Bouzide A (1993) J Chem Soc, Chem Commun 1219
100. Gassman PG, Bottorf KJ (1987) J Am Chem Soc 109: 7547
101. Gassman PG, De Silva SA (1991) J Am Chem Soc 113: 9870
102. Panifilov AK, Cherkaev GV, Magdesieva TV, Prihiyalgovskaya NM (1992) Zh Org Khim 28: 691
103. Adams C, Jacobson N, Utley JHP (1978) J Chem Soc, Perkin Trans 2 1071
104. Minisci F, Citterio A (1983) Acc Chem Res 16: 27
105. Arnoldi C, Citterio A, Minisci F (1983) J Chem Soc, Perkin Trans 2 531
106. Clerici A, Minisci F, Ogawa K, Surzur JM (1978) Tetrahedron Lett 1149
107. Bertrand MP, Surzur JM (1973) Bull Soc Chim Fr 2393
108. Schmittel M (1994) Top Curr Chem 169: 183
109. Schmittel M, Baumann U (1990) Angew Chem Int Ed Eng 29: 541
110. Schmittel M, Abufarag A, Luche O, Levis M (1990) Angew Chem Int Ed Eng 29: 1144
111. Schmittel M, Röck M (1992) Chem Ber 125: 1611
112. Röck M, Schmittel M (1994) J Prakt Chem 336: 325
113. Hoffmann U, Goa Y, Pandey B, Klinge S, Warzecha KD, Krüger C, Roth HD, Demuth M (1993) J Am Chem Soc 115: 10358
114. Ronlan A, Parker VD (1970) J Chem Soc, Chem Commum 1567
115. Ronlan A, Hammerich O, Parker VD (1973) J Am Chem Soc 95: 7132
116. Aalstad B, Ronlan A, Parker VD (1982) Acta Chem Scan B36: 171
117. Taylor EC, Andrade JG, Rall GJH, McKillop A (1980) J Am Chem Soc 102: 6513
118. Taylor EC, Andrade JG, Rall GJH, Turchi IJ, Steliou K, Jagdmann GE, McKillop A (1981) J Am Chem Soc 103: 6856
119. Pandey G (1993) Top Curr Chem 168: 175
120. Pandey G, Krishna A, Madhusudana R (1986) Tetrahedron Lett 27: 4075
121. Pandey G, Krishna A, Bhalerao UT (1989) Tetrahedron Lett 30: 1867
122. Pandey G, Krishna A (1988) J Org Chem 53: 2364
123. Pandey G, Sridar M, Bhalerao UT (1990) Tetrahedron Lett 31: 5373
124. Pandey G, Krishna A, Girija K, Karthikeyan M (1993) Tetrahedron Lett 34: 6631
125. Gilbert BC, McCleland CW (1989) J Chem Soc, Perkin Trans 2 1545
126. Goosen A, McCleland CW, Rinaldi FC (1993) J Chem Soc, Perkin Trans 2 279
127. Lewis FD, Bassani DM, Reddy GD (1992) Pure Appl Chem 64: 1271
128. Lewis FD, Reddy GD (1992) Tetrahedron Lett 33: 4249
129. Lewis FD, Reddy GD, Cohen BE (1994) Tetrahedron Lett 35: 535
130. Katritzky AR, Rees CW (eds) (1984) Comprehensive Heterocyclic Chemistry, vols 1–6. Pergamon Press, Oxford
131. Pandey G, Kumaraswamy G, Reddy PY (1992) Tetrahedron 48: 8295
132. Pandey G (1992) Synlett 546
133. Pandey G, Kumaraswamy G (1988) Tetrahedron Lett 29: 4153
134. Pandey G, Reddy PY, Bhaleao UT (1991) Tetrahedron Lett 32: 5147
135. Cossy J, Guhu M (1994) Tetrahedron Lett 35: 1715
136. Karady S, Corley EG, Abramson NL, Weinstock LM (1989) Tetrahedron Lett 30: 2191
137. Karady S, Corley EG, Abrahamson NL, Amato Js, Weinstock LM (1991) Tetrahedron 47: 757
138. Britcher SF, Lyle TA, Thompson WJ, Varga SL (1988) European Patent No. 0264183
139. Christy ME, Anderson PS, Britcher SF, Colton CD, Evans BE, Remy DC, Engelhart EL (1979) J Org Chem 44: 3117
140. Newcomb M, Deeb TM (1987) J Am Chem Soc 109: 3163
141. Newcomb M, Weber KA (1991) J Org Chem 56: 1309
142. Surzur JM, Stella L, Tordo P. (1970) Bull Soc Chim Fr 115
143. Surzur JM, Stella L, Tordo P. (1970) Tetrahedron Lett 3107
144. Surzur JM, Stella L (1974) Tetrahedron Lett 2191
145. Stella L (1983) Angew Chem Int Ed Engl 22: 337
146. Esker JL, Newcomb M (1993) Adv Heterocycl Chem 58: 1
147. Newcomb M, Marquardt DJ (1989) Heterocycles 28: 129
148. Pandey G, Kumaraswamy G, Bhalerao UT (1989) Tetrahedron Lett 30: 6059
149. Pandey G, Reddy GD, Kumaraswamy G, (1994) Tetrahedron 50: 8185
150. Pandey G, Reddy GD (1992) Tetrahedron Lett 33: 6533
151. Xu W, Zhang XM, Mariano PS (1991) J Am Chem Soc 113: 8863

152. Jung YS, Swartz WH, Xu W, Mariano PS (1992) J Org Chem 57: 6037
153. Kargen H, Lillien I (1966) J Org Chem 31: 3728
154. Crandall JK, Crawley LC (1974) J Org Chem 39: 489
155. Sung HS (1976) J Heterocycl Chem 13: 1351
156. Sung HS (1977) J Heterocycl Chem 14: 693
157. Becker JY, Shakkour E, Sarma JAPR (1990) J Chem Soc, Chem Commun 1016
158. Becker JY, Shakkour E, Sarma JAPR (1992) J Org Chem 57: 3716
159. Engel PS (1980) Chem Rev 80: 99
160. Adam W, Sahin C (1994) Tetrahedron Lett 35: 9027
161. Adam W, Sahin C, Sendelbach J, Walter H, Chen GF, Williams F (1994) J Am Chem Soc 116: 2576
162. Adam W, Sendelbach J (1993) J Org Chem 58: 5310
163. Zona TA, Goodman JL (1993) J Am Chem Soc 115: 4925
164. Engel PS, Robertson JL (1992) Tetrahedron Lett 33: 6093
165. Adam W, Sendelbach J (1993) J Org Chem 58: 5316
166. Shono T, Mitani M (1971) J Am Chem Soc 93: 5284
167. Kariv-Miller E, Mahachi TJ (1986) J Org Chem 51: 1041
168. Swartz JE, Harrold SJ, Kariv-Miller E (1988) J Am Chem Soc 110: 3622
169. Swartz JE, Kariv-Miller E, Harrold SJ (1989) J Am Chem Soc 111: 1211
170. Cossy J, Pete JP, Portella C (1989) Tetrahedron Lett 30: 7361
171. Belotti D, Cossy J, Pete JP, Portella C (1985) Tetrahedron Lett 26: 4591
172. Belotti D, Cossy J, Pete JP, Portella C (1986) J Org Chem 51: 4196
173. Cossy J, Belotti D, Pete JP (1987) Tetrahedron Lett 28: 4545
174. Cossy J, Belotti D, Cuong NK, Chassagnard C (1993) Tetrahedron 49: 7691
175. Cossy J, Madraci A, Pete JP (1994) Tetrahedron Lett 35: 1541
176. Shono T, Nishiguchi I, Ohmizu H (1976) Chem Lett 1233
177. Shono T, Nishiguchi I, Ohmizu H, Mitani M (1978) J Am Chem Soc 100: 545
178. Cossy J, Belotti D, Pete JP (1987) Tetrahedron Lett 28: 4547
179. Cossy J, Belotti D, Pete JP (1990) Tetrahedron 46: 1859
180. Cossy J, Leblanc C (1991) Tetrahedron Lett 32: 3051
181. Cossy J, Belotti D (1988) Tetrahedron Lett 29: 6113
182. Cossy J (1992) Pure Appl Chem 64: 1883
183. Cossy J, Belotti D, Leblanc C (1993) J Org Chem 58: 2351
184. Pattenden G, Teague SJ (1982) Tetrahedron Lett 23: 5471
185. Pattenden G, Robertson GM (1985) Tetrahedron 41: 4001
186. Pradhan SK, Radhakrishnan TV, Subramanian R (1976) J Org Chem 41: 1943
187. Pradhan SK, Kadam SR, Kolhe JN, Radhakrishnan TV, Sohani SV, Thaker VB (1981) J Org Chem 46: 2622
188. Pradhan SK, Kadam SR, Kolhe JN (1981) J Org Chem 46: 2633
189. Kariv-Miller E, Maeda H, Lombardo F (1989) J Org Chem 54: 4022
190. Kariv-Miller E, Maeda H (1992) in Shono T (ed) Modern Methodology in Organic Synthesis, VCH, Weinheim, p 79
191. Kirschberg T, Mattay J (1994) Tetrahedron Lett 35: 7217
192. Lombardo F, Newmark RA, Kariv-Miller E (1991) J Org Chem 56: 2422
193. Fox DP, Little RD, Baizer MM (1985) J Org Chem 50: 2202
194. Little RD, Fox DP, Van Hijfte L, Dannecker R, Sowell G, Wolin RL, Moens L, Baizer MM (1988) J Org Chem 53: 2287
195. Fry AJ, Little RD, Leonetti J (1994) J Org Chem 59: 5017
196. Lee GH, Choi EB, Lee E, Pak CS (1994) J Org Chem 59: 1428
197. Pattenden G, Robertson GM (1983) Tetrahedron Lett 24: 4617
198. Shono T, Kise N (1990) Tetrahedron Lett 31: 1303
199. Shono T, Kise N, Fujimoto T, Tominaga N, Morita H (1992) J Org Chem 57: 7175
200. Shono T, Kise N, Suzumoto T, Morimoto T (1986) J Am Chem Soc 108: 4676
201. Kise N, Suzomoto T, Shono T (1994) J Org Chem 59: 1407
202. Bode HE, Sowell G, Little RD (1990) Tetrahedron Lett 31: 2525
203. Gassman PG, Rasmy OM, Murdock TO, Saito K (1981) J Org Chem 46: 5457
204. Gassman PD, Lee C (1989) Tetrahedron Lett 30: 2175
205. Bischof EW, Mattay J (1990) Tetrahedron Lett 31: 7137
206. Mattay J, Banning A, Bischof EW, Heidbreder A, Runsink J (1992) Chem Ber 125: 2119

207. Pandey G, Hajra S (1994) Angew Chem Int Ed Engl 33: 1169
208. Smith DA, Dimmel DR (1988) J Org Chem 53: 5428
209. Amatore C, Thiebault A, Verpeaux J-N (1989) J Chem Soc, Chem Commun 1543
210. Amatore C, Moustafid TE, Rolando C, Thiebault A, Verpeaux J-N (1991) Tetrahedron 47: 777
211. Gordan M, Ware WR (eds) (1975) The Exiplex. Academic Press, New York
212. Lewis FD, Reddy GD, Schneider S, Gahr M (1989) J Am Chem Soc 111: 6465
213. Lewis FD, Reddy GD (1990) Tetrahedron Lett 31: 5293
214. Lewis FD, Reddy GD, Schneider S, Gahr M (1991) J Am Chem Soc 113: 3498
215. Lewis FD, Bassani DM, Reddy GD (1993) J Org Chem 58: 6390
216. Sugimoto A, Sumida R, Tamai N, Inoue H, Otsuji Y (1981) Bull Chem Soc Jpn 54: 3500
217. Sugimoto A, Yoneda S (1982) J Chem Soc, Chem Commun 376
218. Sugimoto A, Sumi K, Urakawa K, Ikemura M, Sakamoto S, Yoneda S, Otsuji Y (1983) Bull Chem Soc Jpn 56: 3118
219. Sugimoto A, Hiraoka R, Inoue H, Adachi T (1992) J Chem Soc, Perkin Trans 1 1559
220. Mutai K, Yokoyama K, Kanno S, Kobayashi K (1982) Bull Chem Soc Jpn 55: 1112
221. Wubbels GG, Sevetson BR, Sanders H (1989) J Am Chem Soc 111: 1081
222. Mutai K, Kanno S, Kobayashi K (1978) Tetrahedron Lett 1273
223. Mutai K, Kobayashi K, Yokayama K (1984) Tetrahedron 40: 1755
224. Xu W, Jeon YN, Hasegawa E, Yoon US, Mariano PS (1989) J Am Chem Soc 111: 406
225. Yoon UC, Mariano PS (1992) Acc Chem Res 25: 233
226. Xu W, Zhang XM, Mariano PS (1991) J Am Chem Soc 113: 8863
227. Xu W, Mariano PS (1991) J Am Chem Soc 113: 1431
228. Kraus GA, Chen L (1991) Tetrahedron Lett 32: 7151
229. Roth HJ, El Raie MH (1970) Tetrahedron Lett 2445
230. Hasegawa T, Miyata K, Ogawa T, Yoshihara N, Yoshioka M (1985) J Chem Soc, Chem Commun 363
231. Hasegawa T, Miyata K, Ogawa T, Miyata K, Karakizawa A, Komiyama M, Nishizawa K, Yoshioka M (1990) J Chem Soc, Perkin Trans 1 901
232. Hasegawa T, Mukai K, Mizukoshi K, Yoshioka M (1990) Bull Chem Soc Jpn 63: 3348
233. Kanaoka Y (1978) Acc Chem Res 11: 407
234. Mazzocchi PH (1981) Org Photochem 5: 421
235. Coyle JD (1984) in Horspool WA (ed) Synthetic Organic Photochemistry. Plenum Press, New York, p 259
236. Maruyama K, Kubo Y, Machida M, Oda K, Kanaoka Y, Fukuyama K (1978) J Org Chem 43: 2303
237. Machida M, Oda K, Maruyama K, Kubo Y (1980) Heterocycles 14: 779
238. Maruyama K, Kubo Y (1981) J Org Chem 46: 3612
239. Maruyama K, Kubo Y (1978) J Am Chem Soc 100: 7772
240. Kubo Y, Asai N, Araki T (1985) J Org Chem 50: 5484
241. Coyle JD, Newport GL (1977) Tetrahedron Lett 899
242. Roth HJ, Schwartz D, Hundeshagen B (1976) Arch Pharm 309: 48
243. Machida M, Takechi H, Kanaoka Y (1982) Chem Pharm Bull 30: 1579
244. Machida M, Takechi H, Kanaoka Y (1977) Heterocycles 7: 273
245. Mazzocchi PH, Somich C, Edwards M, Morgan T, Ammon HL (1986) J Am Chem Soc 108: 6828
246. Somich C, Mazzocchi PH, Edwards M, Morgan T, Ammon HL (1990) J Org Chem 55: 2624
247. Mazzocchi PH, Somich C (1988) Tetrahedron Lett 29: 513
248. Sato Y, Nakai T, Mizoguchi M, Kawanishi M, Kanaoka Y (1973) Chem Pharm Bull 21: 1164
249. Sato Y, Nakai T, Mizoguchi M, Kawanishi M, Hatanaka Y, Kanaoka Y (1982) Chem Pharm Bull 30: 1263
250. Griesbeck AG, Mauder H, Müller I (1992) Chem Ber 125: 2467
251. Griesbeck AG, Mauder H, Müller I, Peters EM, Peters K, von Schnering HG (1993) Tetrahedron Lett 34: 453
252. Mazzocchi PH, King CR, Ammon HL (1987) Tetrahedron Lett 28: 2473

Received March 1995

One-Electron Redox Reactions between Radicals and Organic Molecules. An Addition/Elimination (Inner-Sphere) Path [1]

Steen Steenken

Max-Planck-Institut für Strahlenchemie, Stiftstrasse 34-36, D-45470 Mülheim, Germany

Table of Contents

In aqueous solution the electron transfer between (reducing) carbon-centered radicals or (oxidizing) hetero-atom-centered inorganic radicals and organic molecules often proceeds by covalent bond formation between the radical and the molecule followed by heterolysis of the so-formed bond between the carbon and the hetero-atom. It is the heterolysis step in which the actual electron transfer between the radical and the molecule takes place. This makes electron transfer a part of the area of (heterolytic) solvolysis reactions. The rate constant for the heterolysis is sensitive to and therefore indicative of the difference in effective electron density or affinity between the radical and the molecule. Factors such as substituents or protonation/deprotonation by which the electron density or distribution is changed strongly influence the rates of heterolysis of the adducts. The observed structure-activity relations for heterolysis of the radical-molecule adducts and thus the electron transfer between the adduct components can be rationalized in terms of the classical solvolysis concepts.

1 Introduction

The pronounced tendency of radicals to engage in one-electron transfer reactions is well documented [3]. This reaction channel is favored because it provides the simplest way for radicals to lose their radical nature, i.e. to become species with an *even* number of electrons (closed-shell molecules). The *direction* of the electron flow between the radical $X^{\bullet}$ and the molecule Y depends on the oxidizing or reducing power of $X^{\bullet}$ and on the ability of Y to either donate or accept an electron: the final result of the interaction between $X^{\bullet}$ and Y is then the either one-electron-reduced or -oxidized former radical (X^- or X^+) or the open-shell molecle ($Y^{\bullet -}$ or $Y^{\bullet +}$), (cf. Eq. 1):

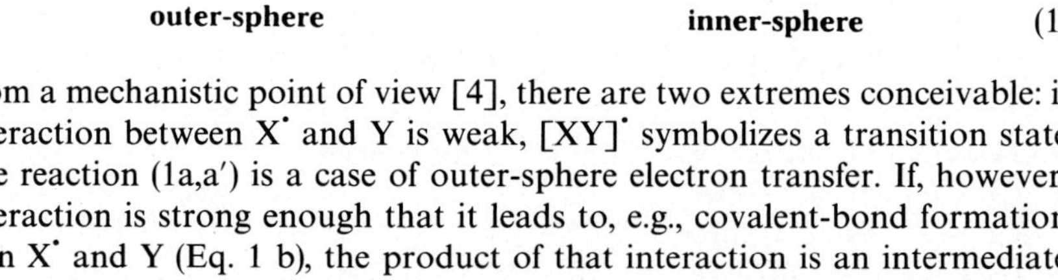

outer-sphere **inner-sphere** (1)

From a mechanistic point of view [4], there are two extremes conceivable: if the interaction between $X^{\bullet}$ and Y is weak, $[XY]^{\bullet}$ symbolizes a transition state and the reaction (1a,a′) is a case of outer-sphere electron transfer. If, however, the interaction is strong enough that it leads to, e.g., covalent-bond formation between $X^{\bullet}$ and Y (Eq. 1 b), the product of that interaction is an intermediate ("adduct") and the overall electron exchange between $X^{\bullet}$ and Y (via Eq. 1b, b′) is then an example of inner-sphere electron transfer.

Equation 1b, b′ is the defining equation for the addition-elimination route for one-electron transfer between $X^{\bullet}$ and Y. It is important to note that although $X{-}Y^{\bullet}$ is a radical and the overall reaction results in the transfer of a *single* electron, in the actual electron transfer step an electron *pair* is shifted rather than a single electron [5]. This means that electron transfer is the consequence of a heterolysis reaction in which the electron pair joining X and $Y^{\bullet}$ ends up at either X or Y$^{\bullet}$, depending on which is the better electrophile. In this sense electron transfer is a sub-area of the general area of (heterolytic) solvolysis reactions, and in parts A and B (vide infra) examples will be presented that tend to support this view.

In addition to the *addition* route Eqs. 1b or 2b, the radicals X–Y$^{\bullet}$ are obtainable by, e.g., an *abstraction* path (Eq. 2a):

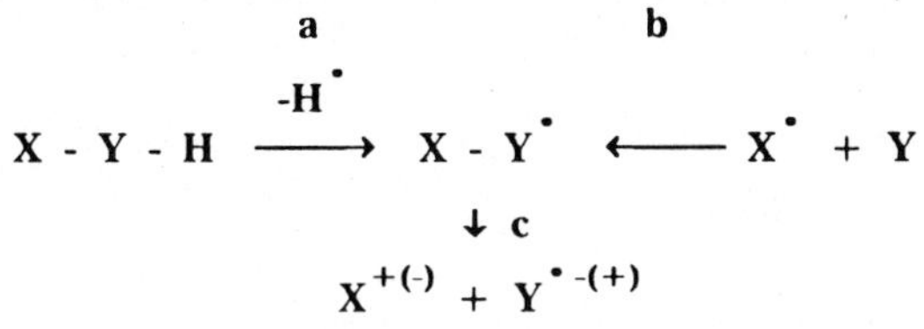

(2)

This enables one to use aliphatic systems as precursors to the radicals X–Y$^{\bullet}$ whose solvolytic (= redox) behavior can then be studied. Equations 2a, c describe what may be called “oxidative solvolysis”. This reaction sequence, the first step of which is in many cases induced by the $^{\bullet}$OH radical, is of great importance in radical (and radiation) chemistry. It extends from β-elimination reactions of monomeric radicals [6, 7] to the mechanism of DNA strand breakage [8]. An example for Eq. 2 in which it is shown that the radical XY$^{\bullet}$ can be produced by either step a or b is given in section 3.3.

The analogy between electron-transfer via addition/elimination (Eq. 2b,c) or abstraction/elimination (Eq. 2a,c) and classical solvolysis involving closed-shell molecules (nonradicals) is seen by comparing Scheme 1 with Scheme 3, in which XY, the precursor of the ions $X^{+(-)}$ and $Y^{-(+)}$, is formally derived from the two *radicals* X$^{\bullet}$ and Y$^{\bullet}$. Analogous to Scheme 1, on the way to the ionic products that result from the interaction between X and Y there are two possibilities: if XY denotes a *transition state*, the reaction (Eq. 3a,a′) is a case of outer-sphere electron transfer. If, however, a covalent bond is formed between X and Y, the path (Eq. 3b,b′) is an example of inner- sphere electron transfer. Obviously, part b′ of the scheme describes the classical area of S_N1 solvolysis reactions (assuming either X or Y to be equal to C) [9, 10]. If a second reaction partner for C (other than the solvent) is allowed for (the (partial) ions then represent transition states), then Eq. 3b′ also covers S_N2 reactions. If looked upon from the point of view of radical-radical reactivity, Eqs. 3a and b show well-known reactions: radical disproportionation in Eq. 3a,a′ and combination in Eq. 3b.

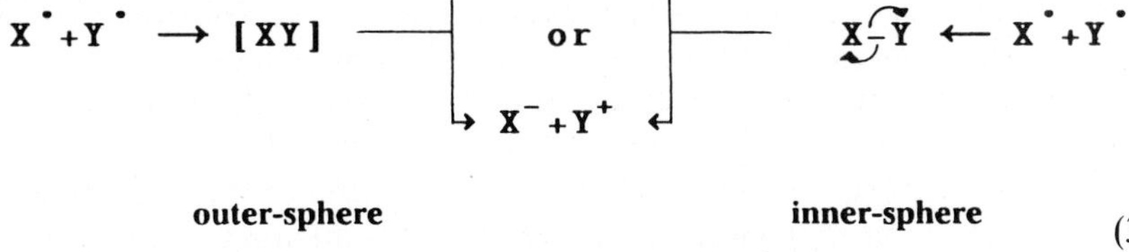

(3)

Concerning the general reaction Scheme 1, attention is restricted to two special areas: A), cases where $X^{\bullet}$ is a carbon-centered radical and Y is an oxygen atom joined by a double bond to some center Z (Eq. 4), and B), cases where $X^{\bullet}$ is a hetero atom, in most cases: oxygen centered radical and Y is a carbon (Eq. 5) [11]. One is then dealing with formation and heterolysis of a bond between a carbon- and a hetero-atom. Of the two, the hetero-atom is of course always more electron-affinic and therefore in the heterolysis the electron pair joining the two will go to the hetero-atom.

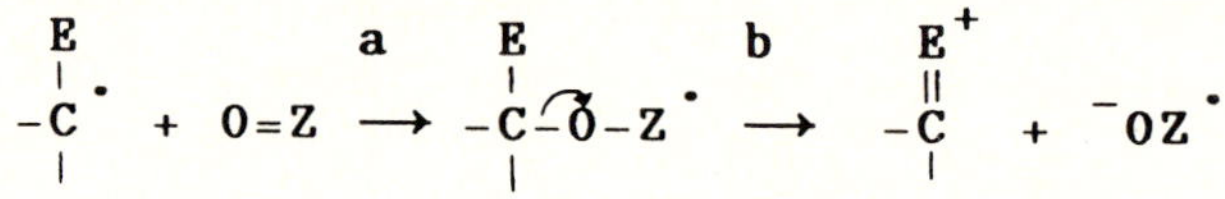
(4)

Z = O, $N(O)C(NO_2)_3$, $N(O)C_6H_4R$, quinone moiety

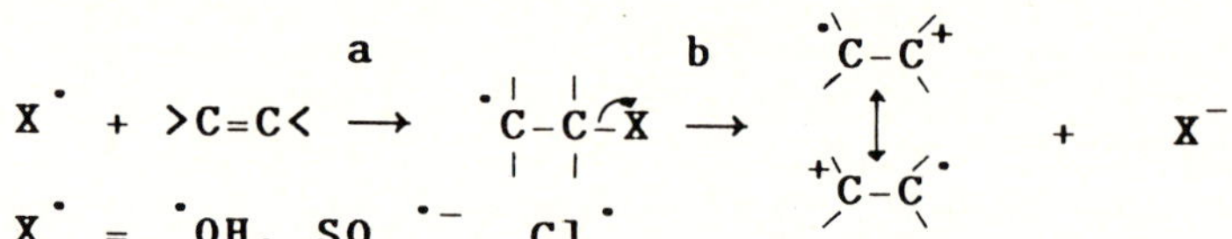
(5)

$X^{\bullet}$ = $^{\bullet}OH$, $SO_4^{\bullet -}$, $Cl^{\bullet}$

The transfer of the electron pair in the heterolysis reaction to the hetero atom (the nucleofugal group) leads to its reduction and to oxidation of the carbon (the electrofugal group). In Eq. 4b the oxidized species is a cation (or its solvolysis products) whereas in the case of Eq. 5b it is a *radical* cation (or its solvolysis products). In the former case the reduced species (the nucleofugal leaving group) is a *radical* anion and in the subsequent Sects. 2 and 3 examples for this type of reaction will be presented. Examples for Eq. 5b, where the nucleofugal leaving group is a closed-shell anion, will be given in Sect. 3.

2 Oxidation of Carbon-Centered Radicals Substituted by a Hetero-Atom E at C_α

2.1 Nitrobenzenes as Oxidants

These reactions belong to the most thoroughly studied ones in the field of radical- molecule reactions in aqueous solution. The interest in this area is to a large part due to its relevance to the understanding of the mechanism of action of nitroaromatics as sensitizers in the radiotherapy of cancer [12].

The hetero atom E at C_α is required for two reasons: (a) to make the addition (of C_α) to the nitro group possible (by providing the necessary "nucleophilicity" [13] to the radical [14], and (b) to stabilize the (incipient) carbocation that results from the heterolysis. The features (a) and (b) are interrelated by the fact that not only the transition state for heterolysis but also that for addition has a large degree of electron transfer character [15].

2.1.1 OH as the Hetero Group at C_α

It was recognized as early as 1968 [16] that the interaction in aqueous solution of these radicals with nitroaromatics can lead to two types of (transient) product: alkoxynitroxyl radicals (produced by addition to the nitro group), and nitro radical anions (cf. Eqs. 6 and 7):

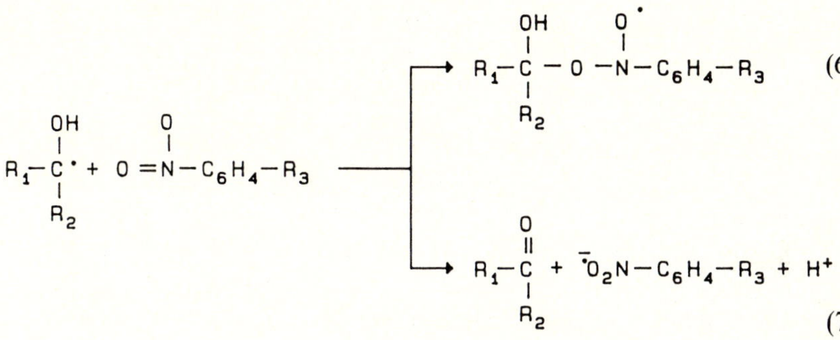

(6)

(7)

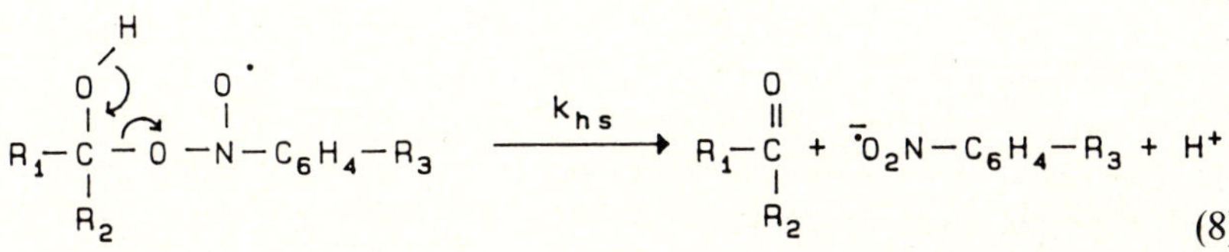

(8)

The tendency to react according to (6) or (7) depends on the stability of the (incipient) carbocation [16, 17] and on the oxidizing power (redox potential) of the nitro compound [12, 18]. It also depends on solvent, more polar solvents favoring the ionic path (Eq. 7) [18].

In aqueous solution the addition (via 6) and the electron transfer product (via 7) do not exclude each other, i.e., (6) and (7) can occur simultaneously [18]. This has been explained in terms of an electron-transfer/addition mechanism involving an ion-pair-type transition state with subsequent competition between combination of the ions (to give addition) and separation by solvent (leading to electron transfer) [15].

The extent to which the radicals react according to Eqs. 6 or 7 depends on the nature of R_1, R_2, and R_3. If $R_1 = R_2 = H$ and $R_3 = H$ through NO_2, the ratio (6):(7) ≥ 20. The addition reactions observed with these systems are characterized by strongly negative activation entropies, which can be rationalized in terms of immobilization of water molecules by the positive charge at C_α in the transition state [15]. That the transition state for addition has pronounced electron-transfer character concluded from the fact [15] that the rate constants for addition depend on the reduction potential of the nitrobenzene in a way describable by the Marcus relation for outer-sphere electron transfer.

The tetrahedral-type alkoxynitroxyl radicals formed in reaction 6 can undergo a C–O heterolysis, (Eq. 8). However, for $R_1 = R_2 = H$ at pH < 6, k_{hs}, the rate constant for spontaneous heterolysis, is only $< 10^2\,s^{-1}$. At pH > 7, a drastic increase of the heterolysis rate occurs, proportional to the concentration of $[OH^-]$. The increase is due to deprotonation of the hemiacetal OH group by which the good electron-donor OH is converted to the excellent one O^-. The electron pair joining C_α and O is thereby "pushed out" leading to k(heterolysis) values of $\geq 5 \times 10^5\,s^{-1}$ [18] (cf. Eq. 9):

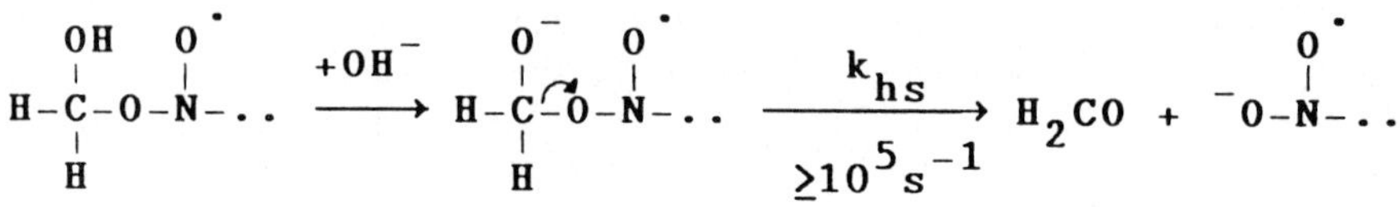

(9)

If one hydrogen at C_α is replaced by the electron-donating methyl group, i.e. $R_1 = H$ and $R_2 = CH_3$, the ratio Eq. (6):(7) = 0.1 to ≈ 3, depending on R_3. Due to the methyl group at C_α, the alkoxynitroxyl radicals are now able to undergo a *spontaneous* heterolysis (Eq. 8) with rate constants ranging from 10^2 to $10^4\,s^{-1}$, depending on R_3. The activation entropies for this process are very negative (− 20 to − 100 $J mol^{-1} K^{-1}$), indicative of the freezing of water molecules in the transition state. As judged by the solvent kinetic isotope effect (k_{H_2O}:k_{D_2O} = 2.2) [18] and by comparison with the much more positive entropy changes observed on heterolysis of nitroxyls from *α-alkoxy*alkyl radicals (see next section), [19] the negative entropy changes are essentially due to deprotonation from the hemiacetal OH.

If $R_1 = R_2 = CH_3$, i.e. with two methyl groups at C_α, the rate of the spontaneous heterolysis (Eq. 8) becomes $\geq 10^6\,s^{-1}$ for all nitrobenzenes with R_3 less electron-donating than CH_3O. This high heterolysis rate constant means that the reaction *appears* to proceed by electron transfer since an intermediate is not visible using detection techniques with μs time resolution. However, the addition/elimination sequence can be demonstrated to exist by using two approaches: (a) reduce the reduction potential of the nitrobenzene (by introducing an electron-donating substituent R_3), or (b) decrease the polarity of the solvent. An example for (a) is the use of the relatively electron-rich 4-nitroaniline which forms a nitroxyl radical with $(CH_3)_2C^{\bullet}OH$ which heterolyzes with $k_{hs} = 2.1 \times 10^3\,s^{-1}$ in H_2O. An example for (b) is the use of 95% *i*-propanol/5% water mixtures. In this solvent the rate of heterolysis of the nitroxyl from $(CH_3)_2C^{\bullet}OH$ and 4-nitrobenzonitrile is slowed down to $1.5 \times 10^4\,s^{-1}$ compared to $\geq 10^6\,s^{-1}$ in pure water [18].

To summarize, by modifying in –C–O– either the leaving group abilities of the carbon moiety (the electrofuge) (e.g., by alkyl substitution at C_α or by ionization of OH) or those of the nitrobenzene (by substitution at the ring) it is possible to go all the way from pure addition to what appears to be pure electron transfer. The heterolysis rates increase with increasing electron "push" by the carbon and with increasing "pull" (= reduction potential) of the nitrobenzene. Also the solvent has a strong influence on the rate constant for heterolysis of the C–O bond, as expected for S_N1-type reactions.

2.1.2 O-Alkyl as the Hetero Group at C_α

On the basis of results obtained from (in-situ-radiolysis [19]) electron spin resonance [16, 20] and pulse radiolysis with optical and conductance detection [19], α-alkoxyalkyl radicals react in aqueous solution exclusively via addition to give alkoxynitroxyl radicals (cf. Eq. 10). This is in contrast to the reactions of $CH_3CH^{\bullet}OH$ (see Sect. 2.1.1) and 5,6-dihydropyrimidine-6-methyl-6-yl radicals (see Sect. 2.1.3), where addition *and* redox products are formed.

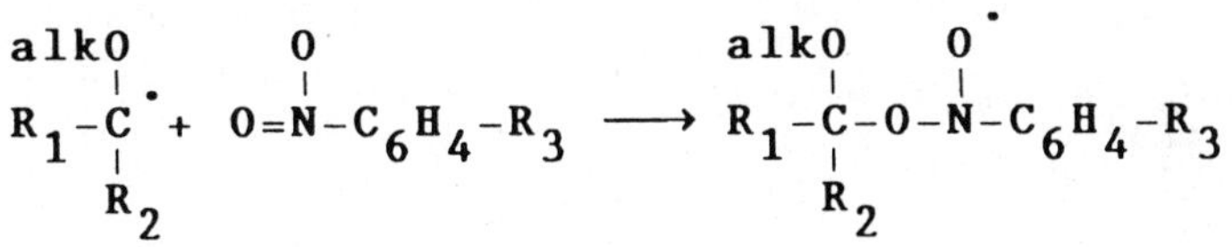

$R_1, R_2 = H, CH_3$, cyclo-alkyl; alk $= CH_3, C_2H_5, (CH_3)_2CH$;

$R_3 = H$ to NO_2 (10)

For the case of $R_3 = COCH_3$, the rate constants for addition increase from e.g. 5.0×10^7 for $R_1 = R_2 = H$ to $1.8 \times 10^9\ M^{-1}\,s^{-1}$ for $R_1 = R_2 = CH_3$. The increase in this direction is due to the activation *entropies* becoming more positive as H is replaced as a substituent at C_α by CH_3. The rate-*enhancing* effect of the more positive activation entropies overcompensates the rate-*decreasing* effect of the higher activation enthalpies that result from the increasing number of methyl groups at C_α [19].

On the basis of the very negative activation entropies, the transition states for the addition are highly ionic, i.e. there is a large degree of electron transfer in the transition state as with the *hydroxy*alkyl radicals (Sect. 2.1.1). In support of this is the fact that the rate constants for addition depend on the reduction potentials of the nitrobenzenes, varied by the substituent R_3 in a way describable by the Marcus equation for outer-sphere electron transfer [19].

For those systems where $R_1 = R_2 = H$ or $R_1 = H$, $R_2 = CH_3$, i.e. where the number of alkyl groups at C_α is ≤ 1, and $R_3 = H$ to NO_3, the alkoxynitroxyl radicals formed according to Eq. 7 under steady-state-ESR or pulse radiolysis conditions do *not* give rise to nitrobenzene radical anions. This means that the rate constants for heterolysis of the nitroxyls are $\leq 10^2\ s^{-1}$. This is not only true in weakly acidic (pH 4) or neutral but also in strongly alkaline solution (pH 13–14). The latter observation means that the nitroxyls are not susceptible to base catalyzed heterolysis. From this the rate constant for OH^- catalyzed decomposition can be estimated to be $\leq 10^2\ M^{-1}s^{-1}$ [19]. This low number for the hypothetical S_N2-type reaction is in line with the known general stability in basic solution of acetal-type compounds. A similar observation was made with acetalic nitroxyl radicals produced by addition of α-alkoxyalkyl radicals to tetranitromethane [21] (see Sect. 2.2). However, these nitroxyl radicals are able to undergo a *spontaneous* heterolysis of the carbon-oxygen bond, a reaction that has all characteristics of S_N1 [21]. The fact that the heterolysis is observable in the case of tetranitromethane as the electron (pair) acceptor and not with the nitrobenzenes is due to the much higher oxidizing or electron deficient character of tetranitromethane as compared to the nitrobenzenes, including 1,4-dinitrobenzene. However, there is one case where heterolysis is observable (by time-resolved optical and conductance methods) also for a nitroxyl formed from a nitrobenzene and an ether radical *mono*-methylated at C_α, (cf. Eq. 11) [19]:

$$CH_3\dot{C}HOC_2H_5 + O_2N{-}C_6H_4{-}N_2^+ \longrightarrow CH_3CH(OC_2H_5){-}O{-}N(O^\bullet){-}C_6H_4{-}N_2^+$$

$$\xrightarrow{+\,H_2O} CH_3CH(OC_2H_5)OH + O_2NC_6H_4^{\bullet} + N_2 + H^+ \quad (11)$$

The reason for the rate increase of the heterolysis from $\leq 10^2\,s^{-1}$ to a range $[(4\text{–}5)\times 10^5\,s^{-1}]$ where it becomes detectable is the exceptionally large electron-accepting power of the N_2^+-substituent (Hammett $\sigma = 1.9$) which results in the electron pair being pulled strongly towards the nitrobenzene.

In order to enhance the C–O heterolysis it is also possible to increase the electron-*donating* properties of the *electro*fugal group (the carbon group) rather than enhance the electron-*attracting* power of the *nucleo*fugal group (the oxygen group). This is most easily done by introducing an additional alkyl group at C_α. In S_N1 reactions of non-radicals an additional methyl group at C_α has been estimated to lead to a rate enhancement by the factor 10^8 [22]. In the nitrobenzene series, spontaneous heterolysis of the alkoxynitroxyl radicals formed by addition of α-alkoxyalkyl radicals is observed for the acyclic system $R_1 = R_2 = CH_3$, alk $= (CH_3)_2CH$ (Eq. 10) and also for the cyclic one:

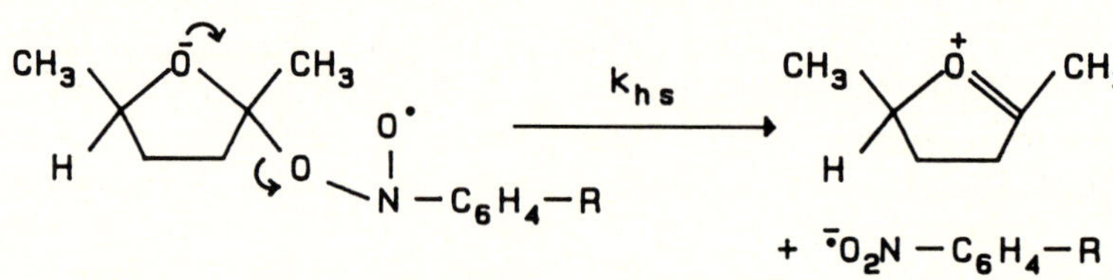

(12)

i.e. for systems with *two* alkyl groups and one alkoxyl group at C_α. The second alkyl group at C_α increases the electron density to an extent which makes the heterolysis rates at 20 °C become of the order $10^5\,s^{-1}$. The rate constants increase with increasing electron-deficiency of the substituent R on the nitrobenzene, i.e. with increasing electron pull by the electron acceptor. The heterolysis reaction is characterized by activation enthalpies of $\approx 50\,kJ\,mol^{-1}$ and by *positive* activation entropies, $\approx 20\,J\,mol^{-1}\,K^{-1}$, which is in contrast to the case of the heterolysis of the nitroxyls formed by addition of a *hydroxy*ethyl radical to the same nitrobenzenes, where the activation entropies are strongly negative (to $-100\,Jmol^{-1}\,K^{-1}$) [18]. The difference between the two systems is probably due to the fact that in the case of the hydroxyethyl nitroxyls deprotonation from the hemiacetal OH group takes place in the transition state with hydration of the (incipient) proton thereby leading to freezing of water molecules which causes the loss of entropy, i.e.

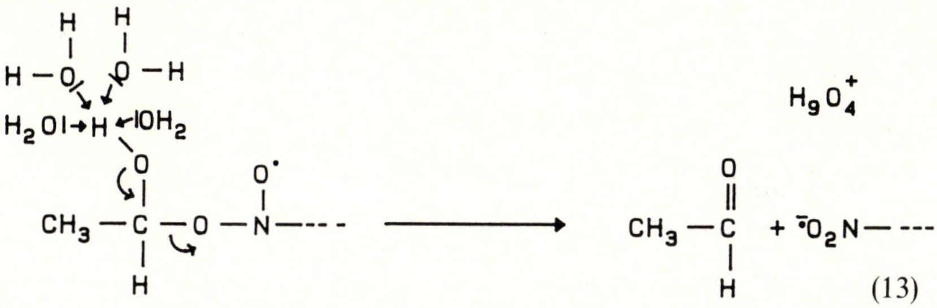

(13)

This means that C–O bond breaking is to a certain degree concerted with O–H bond breaking, or, expressed differently, solvent assisted O–H deprotonation provides the necessary electron density at C_α to make the C–O heterolysis possible. In agreement with this picture is the fact that replacement of the *hemi*acetal H by an alkyl group (i.e. going to the C_α acetalic systems) makes the rate constant for heterolysis drop below $10^2\,s^{-1}$. Only after introducing an additional CH_3 group at C_α with the concomitant increase in electron density does the heterolysis become detectable again ($k_{hs} \approx 10^5\,s^{-1}$).

2.1.3 –N(CO)– as the Hetero Group at C_α (5,6-Dihydropyrimidine-6-yl Radicals)

The 6-yl radicals produced by (the selective) [23, 24] addition of $^{\bullet}OH$ to C(5) of the C(5)/(6) double bond of naturally occurring pyrimidine bases, nucleosides and nucleotides or those formed by H-abstraction from C(6) of 5,6-dihydropyrimidines [25] react with para-substituted nitrobenzenes by addition ($k \approx 6 \times 10^6$ to $2 \times 10^9\,M^{-1}\,s^{-1}$) to yield nitroxyl-type radicals which were identified by ESR and optical detection techniques [26, 27] (cf. Eq. 14):

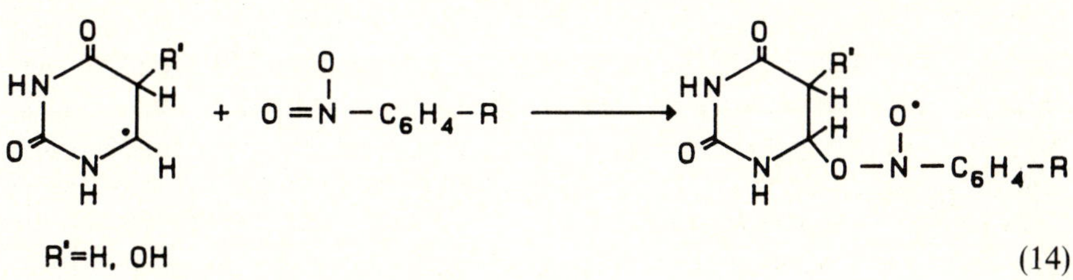

(14)

At pH < 7 the nitroxyl radicals do not undergo an observable heterolysis ($k_{hs} \leq 10^2\,s^{-1}$), but decay by bimolecular reactions. However, in basic solution an OH^--catalyzed heterolysis takes place to yield the radical anion of the nitrobenzene and an oxidized pyrimidine. In the case of the nitroxyls substituted at N(1) by H (i.e. those derived from the free bases), the OH^- catalysis involves deprotonation at N(1) which is adjacent to the reaction site [= C(6)] (cf. Eq. 15) [26]:

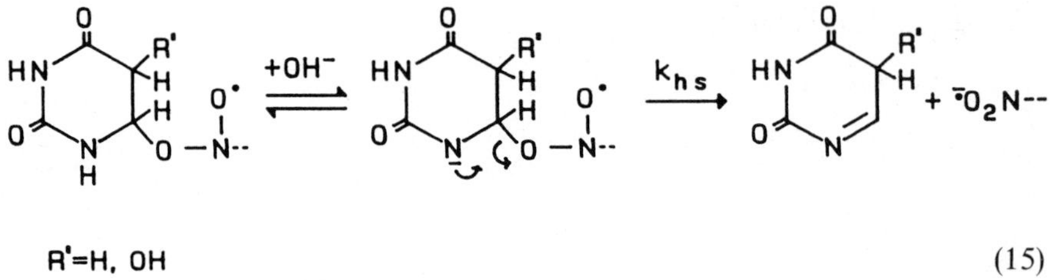

(15)

Deprotonation provides the necessary electron push to kick out the electron pair joining C(6) with the nitrobenzene oxygen. If, however, N(1) is alkylated (as with the nucleosides and nucleotides), OH^- catalysis is much less efficient since it now proceeds by deprotonation from N(3) (with the uracils) or from the amino group at C(4) (with the cytosines). In these cases the area of deprotonation is separated from the reaction site by a (hydroxy)methylene group which means that the increase in electron density that results from deprotonation at N(3) is transferable to the reaction site only through the carbon skeleton (inductive effect), which is of course inefficient as compared to the electron-pair donation from N(1) (mesomeric effect) [26]. Reaction 15 is a 1:1 model for the catalytic effect of OH^- on the heterolysis of peroxyl radicals from pyrimidine-6-yl radicals (see Sect. 2.4).

The ionized 6-yl radicals react with the nitrobenzenes also by addition, however with rate constants considerably higher ($k = 1 \times 10^8$ to $1 \times 10^9\,M^{-1}\,s^{-1}$) than those for the case of the neutral radicals. This indicates that also the transition state for the *addition* reaction is ionic [26]. The same conclusion can be reached from the increase of the rate constants for addition with increasing reduction potential of the nitrobenzenes and also from the very negative activation entropies for the addition reaction [15].

The 5,6-dihydropyrimidine-6-yl radicals discussed above behave, in their reactions with nitrobenzenes, like the simpler radicals $^{\bullet}CH_2OH$ and $^{\bullet}CH$(alkyl)Oalkyl do, i.e. they react exclusively by addition to give nitroxyl radicals and uncatalyzed heterolysis is not observed ($k_{hs} \leq 10^2\,s^{-1}$). If, however, a methyl group is introduced at C(6) (= C_α) of the pyrimidine-6-yl radical, the corresponding nitroxyl radicals heterolyze with rate constants at 20 °C of 10^3 to $5 \times 10^5\,s^{-1}$ depending on the structure of the pyrimidine and of the nitrobenzene (Eq. 16). This S_N1 type reaction is characterized by activation enthalpies of 30–40 $kJ\,mol^{-1}$ and activation entropies of -89 to $-7\,J\,mol^{-1}\,K^{-1}$ (entropy control) [27]. The rate-enhancing effect of the methyl group is, of course, due to stabilization of the (incipient) C(6) carbocation developing in the C–O heterolysis. The rate constants for the heterolysis reaction are a measure of the reducing power of 5,6-dihydro-6-methylpyrimidine-6-yl radicals. On this basis, the cytosine radicals are better reductants than the corresponding uracil radicals, and the radicals derived by hydrogen atom addition to pyrimidines are stronger reductants than those formed by OH radical addition [27].

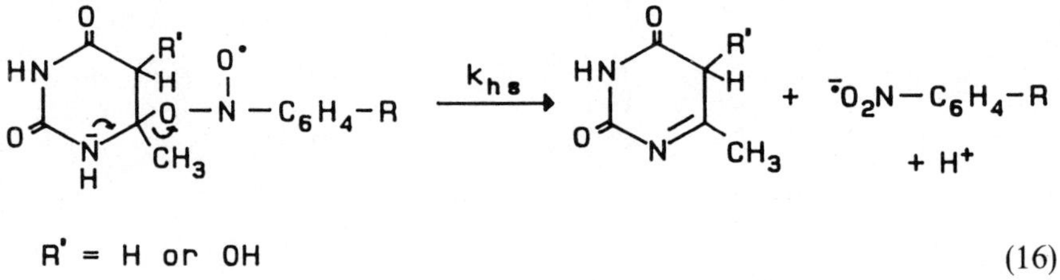

(16)

2.2 Tetranitromethane as an Oxidant

It was mentioned above that acetalic nitroxyl radicals produced by addition of α-alkoxyalkyl radicals to tetranitromethane (TNM) undergo a spontaneous heterolysis with the carbon center being oxidized and TNM reduced (to nitroform anion (NF^-) and $NO_2^{\bullet}$). In order to *see* the addition-elimination sequence with acyclic α-alkoxyalkyl radicals there have to be two (electron-withdrawing relative to methyl) *hydrogens* at C_α. Even one alkyl group at C_α is sufficient to make $k_{hs} \geq 10^6\ s^{-1}$ and therefore too fast to measure. If, however, an alkyl group which is inductively *de*activated is introduced at C_α, the k values fall in the experimentally accessible range (10^2–$10^6\ s^{-1}$). An example for this is the dioxolan-3-yl/TNM system (cf. Eq. 17):

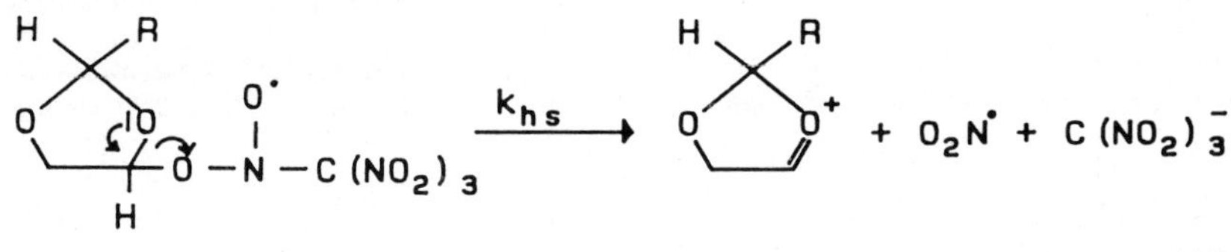

(17)

By varying R at C_γ (which is separated from the reaction site at C_α by an oxygen) it is possible to influence the electron density at C_α in a defined way to change the heterolysis rate constant. Corrected for the incomplete transmission of electronic effects through the C methylene group and assuming that the transmission of the oxygen is 100%, the Taft ρ^* value for the effect of substitution on the heterolysis rate constant is equal to -3.95, comparable to that ($\rho^* = -3.29$) for the S_N1 hydrolysis of tertiary alkyl halides [21].

That the heterolysis has pure S_N1 character was also deduced from the activation parameters of the heterolysis and from its solvent dependence [21].

2.3 Anthraquinone-2,6-Disulfonate as an Oxidant

This quinone reacts in aqueous solution with OH and H adducts of cytosines and uracils by an electron transfer/addition mechanism, similar to Eq. 18 [28]. Addition takes place at the quinone carbonyl oxygen to produce an anthroxyl radical. This then undergoes spontaneous C–O heterolysis:

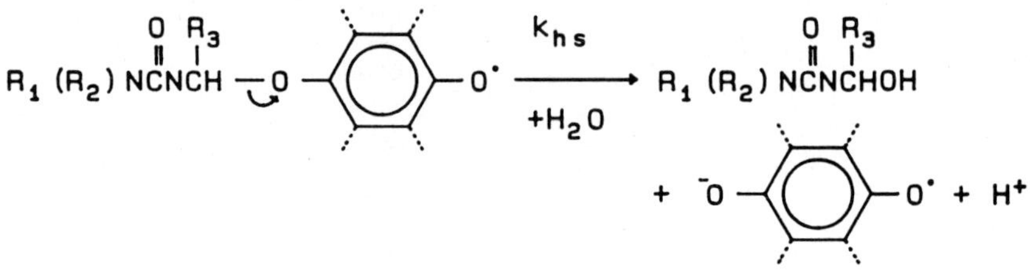

(18)

The activation parameters for the (bimolecular) addition and the (unimolecular) heterolysis steps have been determined [28] for the case of R_1, R_2, $R_3 = H$ or CH_3 and the results are shown in Fig. 1. It is obvious that the heterolysis reaction is entropy controlled which is the consequence of the highly ionic transition state which leads to freezing of water molecules with the concomitant loss of entropy.

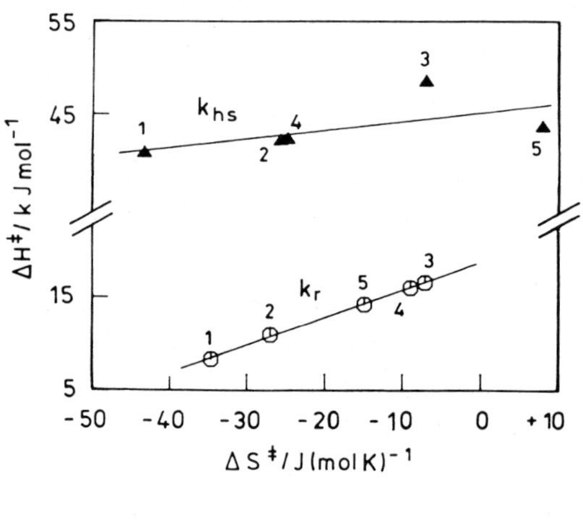

Fig. 1. Isokinetic plot: The dependence on R of the activation parameters for formation in H_2O, k_r, and those for heterolysis (acc. to Eq. 18) of the anthroxyl radical, k_{hs}. Key: 1: $R_1 = R_2 = R_3 = H$. 2 and 3: $R_1 = R_2 = H$, $R_3 = CH_3$. 3: solvent D_2O. 4: $R_1 = R_3 = CH_3$, $R_2 = H$. 5: $R_1 = R_2 = R_3 = CH_3$

2.4 O_2 as an Oxidant

The way O_2 reacts with many radicals from biologically important precursors is quite similar to that described in sections 2.1–2.3 for nitro compounds and the quinone. It is for this reason that quinones and nitro compounds can to a certain degree replace O_2 as a radiation sensitizer. O_2 reacts with organic radicals by *addition* [29], and this in spite of the fact that electron transfer is often thermodynamically highly favorable. (For example, in the case of the reaction of O_2 ($E_7 = -0.155$ V/NHE) with $(CH_3)_2C^{\bullet}OH$ ($E_7 = -2.2$ V/NHE) [30], the difference in the reduction potentials of the reactants (≈ 2 V) corresponds to a driving force of $\approx 46\,kcal\,mol^{-1}$ for the *hypothetical* outer-sphere electron transfer according to Eq. 19:

$$(CH_3)_2\dot{C}OH + O_2 \not\rightarrow (CH_3)_2CO + O_2^{\bullet -} + H^+ \tag{19}$$

However, in spite of this enormous exothermicity, the reaction does *not* proceed by electron transfer but by addition [31], cf. Eq. 20a,

$$(CH_3)_2\dot{C}OH + O_2 \xrightarrow{a} (CH_3)_2C(OH){-}O_2^{\bullet} \xrightarrow{b} (CH_3)_2CO + O_2^{\bullet -} + H^+ \tag{20}$$

and the adduct (the hydroxyperoxyl radical) then undergoes spontaneous C–O heterolysis ($k_{hs} = 670\,s^{-1}$) (cf. Eq. 20b) to give the oxidized organic, i.e. acetone, and the reduced O_2, i.e. $O_2^{\bullet -}$ [31]. The hydroxyperoxyl radical is of the hemiacetal type, and base catalysis of its heterolytic decomposition is therefore possible. It involves deprotonation of the OH group to give the substituent $-O^-$ whose greatly increased electron density (as compared to –OH) increases the heterolysis rate to $> 10^5\,s^{-1}$ [31, 32].

The heterolysis rate is also increased by introducing a second OH [33] or Oalkyl [34] group al C_α, e.g.

$$HC(OH)_2{-}O_2^{\bullet} \longrightarrow HC(O)OH + O_2^{\bullet -} + H^+;\ k_{hs} \geq 10^6\ s^{-1}. \tag{21}$$

The reason for the rate enhancement compared to Eq. 20 is the added stabilization of the (incipient) carbocation by the *two* OH groups at C_α.

At this stage it may be interesting to compare the one-electron oxidizing efficiency of O_2 with that of, e.g., nitrobenzene, taking $(CH_3)_2C^{\bullet}OH$ as the common reductant. In aqueous solution the rate constant for heterolysis of the adduct of $(CH_3)_2C^{\bullet}OH$ to the *weak* oxidant nitrobenzene ($E_7 = -0.486$ V/NHE) [30] is $\geq 10^6\,s^{-1}$ [18], whereas that for heterolysis (see Eq. 20b) of the adduct of $(CH_3)_2C^{\bullet}OH$ to the *strong* oxidant O_2 ($E_7 = -0.155$ V/NHE) [30] is only $670\,s^{-1}$ [31]. The weak oxidant nitrobenzene thus leads to the product (acetone) considerably faster than does the strong oxidant O_2. This, at first sight puzzling, result can be understood if the leaving group abilities of the reduced oxidants are considered. On the basis of the pK_a values of the conjugate acids, the nitrobenzene radical anion (pK = 3.2) [35] is a better leaving group than $O_2^{\bullet -}$ (pK = 4.8) [36].

A further example for oxidation by O_2 are 5,6-dihydropyrimidine-6-yl radicals. These react with O_2 to give the 6-peroxyl radicals shown [23, 24, 25, 37].

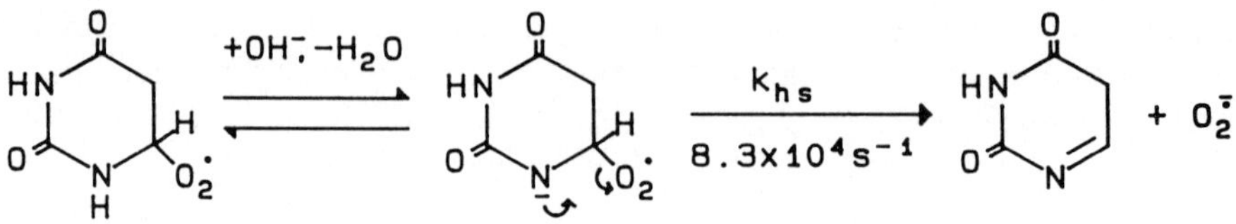

(22)

With these radicals, spontaneous C(6)–O heterolysis is slow ($< 10^3\,s^{-1}$). However, if the electron density of the system is increased by OH^--induced deprotonation of N(1)-H, $O_2^{\bullet -}$ elimination is observed [23, 24, 25]. With the peroxyl radical from 5,6-dihydrouracil-6-yl, the heterolysis rate constant is $8.3 \times 10^4\,s^{-1}$, the reaction leading to the isopyrimidine derivative shown [37]. The reaction is perfectly analogous to the eliminations of the radical anions of nitrobenzenes (Eq. 15) or anthraquinone-2,6-disulfonate (Eq. 18).

3 Oxidation of Olefinic and Aromatic Compounds

3.1 $^{\bullet}$OH as the Oxidant [38]

Thermodynamically, the OH radical is a very powerful one-electron oxidant. Its reduction potential at pH 0 is 2.7 V/NHE [39], and at pH 7 it is still 2.3 V. That this number indicates strong oxidizing power is evident on comparing it with those of some well known oxidants such as $IrCl_6^{2-}$ (0.87 V) or Tl^{2+} (2.2 V). In spite of this, $^{\bullet}$OH does usually not react by electron transfer but by addition, not only with organic substrates (containing double bonds), but also with anions [40] and even metal ions [41]. This tendency to add rather than to oxidize is probably caused by stabilization of the transition state for addition by contributions from bond *making*, whereas electron transfer requires pronounced bond and solvent reorganization with a correspondingly large free energy change to reach the transition state.

$^{\bullet}$OH addition leads to the "OH adduct" HO–Y$^{\bullet}$ (Eq. 23). In order for this adduct to yield electron transfer products, heterolysis of the bond joining HO and Y has to occur. However, due to the fact that OH^- is a very bad leaving group (as evidenced by the high pK_a (15.7) of its conjugate acid, H_2O), the rate of the spontaneous heterolysis, k_{hs} (Eq. 23a), is very low (very often, $\ll 10^2\,s^{-1}$). As a consequence, the final (non-radical) products from $^{\bullet}$OH reactions with Y are typically derived from dimerization or disproportionation of HOY$^{\bullet}$.

One-electron oxidation of Y by $^{\bullet}$OH is, however possible by changing the leaving group abilities of the adduct components, HO– and Y$^{\bullet}$. As shown in

Eq. 23b, protonation of HO^- to give H_2O^+-converts the bad leaving group OH^- into the excellent one H_2O ($pK_a(H_3O^+) = -1.7$). If it is assumed that the Brönsted catalysis law is applicable to this case and the Brönsted coefficient, α, is equal to 0.5, a rate enhancement of $10^{8.7}$ induced by protonation of the leaving group –OH is calculated from the difference in the pK_a values of H_2O and H_3O^+. A somewhat similar number (10^7) is obtained[3d] by considering the difference in reduction potential of $^{\bullet}OH$ at pH 7 and at pH 0.

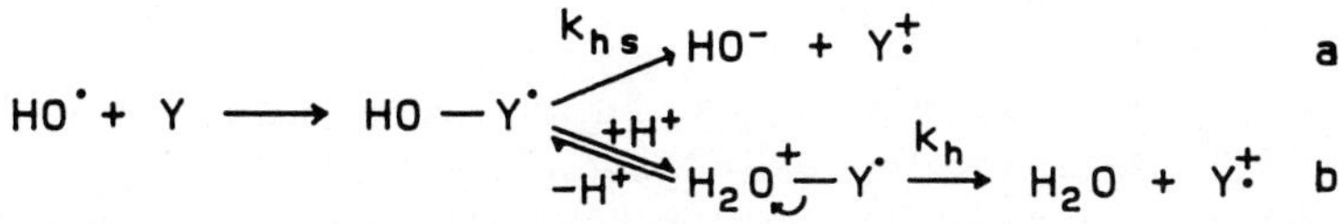

(23)

Protonation is a very effective method to improve the *nucleo*fugacity of the leaving group OH. It results in an increase in the rate of heterolysis of the HO–$Y^{\bullet}$ bond. The reciprocal way, which can be even more efficient, is to improve the *electro*fugacity of $Y^{\bullet}$. A way to achieve this is to introduce electron-donating substituents into $Y^{\bullet}$. An elegant (and important) method is to increase the electron density on $Y^{\bullet}$ by ionization of a substituent which is a Brönsted acid. An example for this, which serves also to summarize the mechanisms of H^+- and OH^--supported dehydration, is shown in Eq. 24.

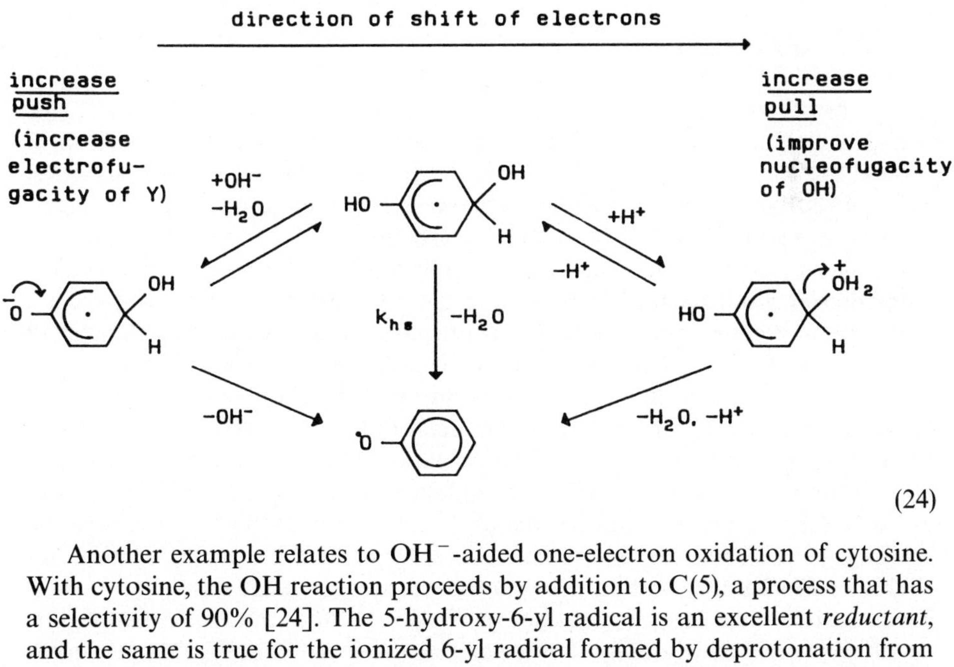

(24)

Another example relates to OH^--aided one-electron oxidation of cytosine. With cytosine, the OH reaction proceeds by addition to C(5), a process that has a selectivity of 90% [24]. The 5-hydroxy-6-yl radical is an excellent *reductant*, and the same is true for the ionized 6-yl radical formed by deprotonation from N(1). This radical anion now contains sufficient electron density to eliminate the OH group at C(5) as OH^-. The result is the cytosine-1-yl radical which is *oxidizing*, probably due to appreciable spin density at the hetero atoms N(1) and

O^2 [24].

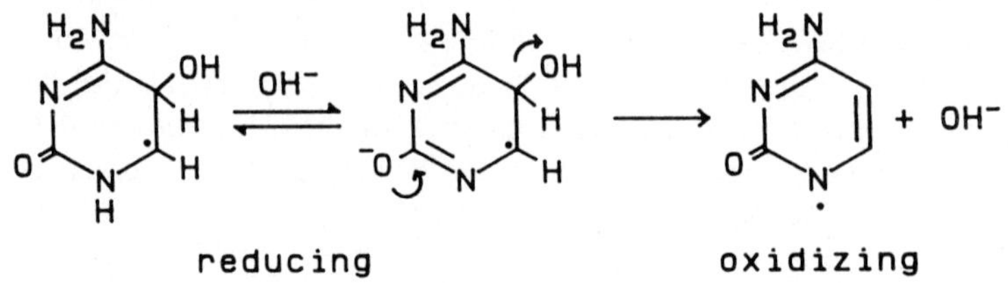

(25)

Reactions (24) and (25) are examples for the general phenomenon of "redox inversion" [38] by dehydration of OH adducts.

The reaction between $^{\bullet}OH$ and phenol lends itself to an analysis of its thermochemistry. On the basis of $E_7(^{\bullet}OH) = 2.3$ V/NHE and $E_7(PhO^{\bullet})$ $= 0.97$ V/NHE [42], the formation of $PhO^{\bullet}$ and H_2O via an electron-transfer mechanism is exothermic by 1.33 V = 31 kcal mol^{-1}. In spite of this, the reaction proceeds by addition, as outlined in Eq. 24. Again, the propensity of OH to add rather than to oxidize can be understood in terms of the transition state for addition being stabilized by contributions from bond *making*, in contrast to electron transfer which requires pronounced bond and solvent reorganization which results in a large (entropy-caused) free energy change.

3.2 $SO_4^{\bullet-}$ as an Oxidant

This radical, which, like $^{\bullet}OH$, is a strong one-electron oxidant (E 2.5–3.1 V/NHE) [4c], has been proposed to react by outer-sphere electron transfer with aromatics [43]. However, on the basis of a Marcus treatment considering the redox potentials of the species involved, some scepticism concerning this mechanism is appropriate [4c]. With certain simple alkenes $SO_4^{\bullet-}$ reacts by addition [44], whereas with alkenes carrying (electron-donating) alkyl groups, one-electron-transfer products have been observed [44b]. Electron transfer products are also seen on reaction with electron-rich benzenes [45]. The question thus arises whether there is a duality of reaction mechanism or whether the electron-transfer products are the result of an addition/elimination sequence [6a.44b]. Using the method of time-resolved conductance and appropriately substituted alkenes it is possible to show that their oxidation proceeds by addition followed by heterolysis of the so-formed SO_4^- adduct [2]. For instance, in the case of cyclohexene in aqueous solution, the SO_4^- adduct formed in the addition step hydrolyzes at 20 °C with a rate constant $k = 3.0 \times 10^4\ s^{-1}$, an activation enthalpy of 17 kJ mol^{-1} and an activation entropy of -103 J mol^{-1} K^{-1}) [46]. This negative value for a decomposition reaction (with an intrinsic entropy gain) can be explained by assuming that water molecules are immobilized in the ionic heterolysis transition state. The reaction is of the S_N1 type, a conclusion that is supported by (a) the lack of effect of the strong nucleophile OH^- on the heterolysis rate and, (b) the strong effect of methyl substitution at C_α (the electrofuge): in the case of 1-methylcyclohexene-1 the SO_4^- adduct heterolyzes with $k_{hs} \geq 5 \times 10^6 \, s^{-1}$ [46].

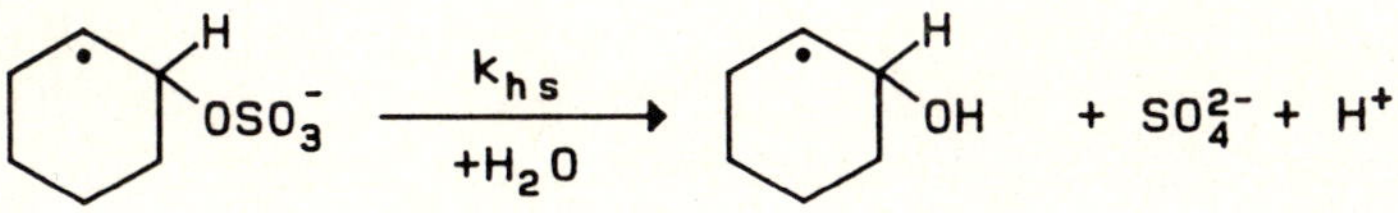

(26)

The addition/elimination type of reaction that $SO_4^{\bullet -}$ undergoes with alkenes may be compared with the pure addition behavior of $^{\bullet}OH$. The reduction potentials of the two radicals being similar means that their oxidizing power is comparable. The difference between the two thus lies in the *leaving group abilities* of their redox partners, i.e. SO_4^{2} and OH^-, respectively. SO_4^{2-} is a much better leaving group than OH^-. On the assumption that the Brönsted catalysis law is valid in this case and that Brönsted $\alpha = 0.5$, the difference between the pK_a values of the conjugate acids, HSO_4^- ($pK_a = 1.9$) and H_2O ($pK_a = 15.7$) translates into a difference in the rates of heterolysis of SO_4^- and OH adducts corresponding to a factor of $10^{6.9}$. Such a large factor, of course, makes $^{\bullet}OH$ appear to react in an altogether different way as compared to $SO_4^{\bullet -}$.

In this connection it is interesting to compare the leaving group abilities of SO_4^{2-} and H_2O. On the basis of the pK values (pK_a ($H_3O^+ = -1.7$) and again assuming that Brönsted $\alpha = 0.5$, H_2O is a better leaving group than SO_4^{2-} by the factor $10^{1.8}$. This explains why in strongly acidic solutions OH behaves as an (apparent one-electron) oxidant [3d, 6, 44b, 45] of strength comparable to that of $SO_4^{\bullet -}$.

Concerning $SO_4^{\bullet -}$ reactions with aromatics, adducts have so far not been identified. For instance, if an SO_4^- adduct is formed with benzene, it hydrolyzes to give the hydroxycyclohexadienyl radical with a rate constant of $> 10^7 \, s^{-1}$ [43]. Even if the strongly electron withdrawing CN group is introduced into the benzene moiety to reduce its electrofugal leaving group ability (i.e. the benzonitrile/$SO_4^{\bullet -}$ system), the rate constant for heterolysis is still $> 5 \times 10^6 \, s^{-1}$ [47].

3.3 $Cl_2^{\bullet -}$ as an Oxidant

This radical has been shown [48] to react by oxidation or by addition, depending on the nature of the substrate. The reduction potential in aqueous solution, E ($Cl_2^{\bullet -} + e^- \rightarrow 2\,Cl^-$) is equal to 2.1 V/NHE, i.e. less than that of $^{\bullet}OH$ or $SO_4^{\bullet -}$. Since the reduction potential of the related system, $E(Cl^{\bullet} + e^- \rightarrow Cl^-) = 2.4$ V/NHE is less than those of $^{\bullet}OH$ and $SO_4^{\bullet -}$, it is thermodynamically possible to oxidize Cl^- with these radicals to give $Cl^{\bullet}$, and subsequently, by reaction of $Cl^{\bullet}$ with Cl^-, $Cl_2^{\bullet -}$ is obtained ($k(Cl^{\bullet} + Cl^-) = 2.1 \times 10^{10} \, M^{-1} s^{-1}$) [44a, 49]. It may be interesting at this point to recall that the oxidation by $^{\bullet}OH$ of Cl^- (and other halides and pseudohalides) proceeds by an addition/elimination sequence [40] with H^+ serving to convert the bad leaving group OH^- into the good one H_2O, as outlined for the general case in Scheme 23.

The rate constant for reaction of $SO_4^{\bullet -}$ with Cl^- is $3.1 \times 10^8\ M^{-1}\,s^{-1}$ [44a]. By introducing an excess of Cl^- over that of an organic substrate in a system in which $SO_4^{\bullet -}$ radicals are produced it is thus possible to produce preferentially $Cl_2^{\bullet -}$ and make it react with the organic [48].

If an experiment of this type is performed with an aqueous solution saturated with isobutene and containing 20 mM Cl^-, a unimolecular rise of conductance of the solution occurs after production (< 1 µs) of $SO_4^{\bullet -}$ radicals. At 20 °C the rate constant of this conductance change, which is independent of pH between 4 and 11, is $3.1 \times 10^4\ s^{-1}$ [46]. These results are explained by reactions (27)–(30), (29) and (30) constituting the actual addition/elimination sequence:

$$SO_4^{\bullet -} + Cl^- \longrightarrow SO_4^{2-} + Cl^{\bullet} \tag{27}$$

$$Cl^{\bullet} + Cl^- \longrightarrow Cl_2^{\bullet -} \tag{28}$$

$$Cl_2^{\bullet -} + (CH_3)_2C{=}CH_2 \longrightarrow (CH_3)_2\dot{C}CH_2Cl + Cl^- \tag{29}$$

$$(CH_3)_2\dot{C}CH_2Cl + H_2O \xrightarrow{k_{hs}} (CH_3)_2\dot{C}CH_2OH + Cl^- + H^+ \tag{30}$$

The identification of reaction (30) as that causing the first-order-conductance increase is supported by the fact that the same first-order conductance change is observed on producing $(CH_3)_2C^{\bullet}CH_2Cl$ by H abstraction by $^{\bullet}OH$ from isobutylchloride (Eq. 31):

$$(CH_3)_2CHCH_2Cl + \dot{O}H \rightarrow (CH_3)_2\dot{C}CH_2Cl + H_2O \tag{31}$$

The oxidative solvolysis steps (Eq. 31/30) have previously been demonstrated to occur, and the room temperature rate constant given ($k_{hs} = 3.5 \times 10^4\ s^{-1}$)[7] is in good agreement with that measured on producing the β-chloroalkyl radical by the addition route (Eq. 29).

A further support for the identification of the species responsible for the unimolecular conductance increase in terms of a chlorine-containing radical is the fact that in a blank experiment, i.e. one in which chloride is left out, a slow conductance increase is *not* observed and the overall conductance yield is only half of that in the *presence* of chloride. Since in isobutene-saturated aqueous solution the lifetime of $SO_4^{\bullet -}$ is only 90 ns due to its rapid reaction with the alkene (as determined by optical experiments at 450 nm) [46], the non-observability of a unimolecular conductance increase means that the rate constant for heterolysis of the $SO_4^{\bullet -}$ adduct to isobutene is $\leq 10^2\ s^{-1}$ (cf. Eq. 32):

$$(CH_3)_2\dot{C}CH_2OSO_3^- + H_2O \rightarrow (CH_3)_2\dot{C}CH_2OH + H^+ + SO_4^{2-};\ k_{hs} \leq 10^2\ s^{-1} \tag{32}$$

It is thus seen that in the reaction with isobutene the weaker oxidant $Cl_2^{\bullet -}$ gives rise to the product $(CH_3)_2C^{\bullet}CH_2OH$ (the oxidized alkene) with a considerably faster rate than does the stronger oxidant $SO_4^{\bullet -}$. It is thus clear that in this sense the activity of the oxidant is *not* determined by its oxidizing power but by the leaving group quality of the conjugate redox partner: Obviously, Cl^- is a much better leaving group (pK_a (HCl) = − 7) than SO_4^{2-} (pK_a (HSO_4^-) = 1.9). That Cl^- is an excellent leaving group also in the solvolyses of other radicals is well documented [7].

4 Summary and Conclusions

1. Examples have been given for one-electron redox reactions between organic molecules and reducing or oxidizing organic or inorganic radicals, reactions that proceed by addition followed by elimination whereby the process separating the components of the adducts is a heterolysis involving a carbon and a hetero atom. Addition takes place even in cases where electron transfer is strongly exothermic. The reason for this preference for addition is that the transition state for addition is lower in energy than that for electron transfer because addition profits from bond making, whereas electron transfer requires entropically expensive bond and solvent reorganization [4].
2. The efficiency of an oxidant or a reductant does not necessarily depend solely on its redox potential. It is a characteristic of any inner-sphere or "bonded" [50] electron transfer mechanism that the *leaving group properties* of the conjugate redox partner, i.e. the reduced oxidant or the oxidized reductant, respectively, are of great importance in determining the overall efficiency of the oxidant or the reductant. The leaving group properties are not necessarily related in a simple way to the reduction potentials. This is clear if Eqs. 33 and 34 are compared. Equation 33 defines the leaving group properties: If Y = H, the reaction defines the Brönsted acidity (of X–H) or basicity (of X^-) (Eq. 33a). If Y = –C<, the electro- or nucleo*fugacities* and the electro- or nucleo*philicities* are defined (eq. 33b). Obviously (a) and (b) are related since they differ only by the electrofuge (H vs –C<). The reaction (= heterolysis [5]) consists in a shift of an electron pair between X and Y with more (S_N2) or less (S_N1) participation of a second reaction partner (other than the solvent). The common feature is the propensity of X to accept the bonding electron pair joining Y with X.

$$Y - X \rightleftarrows Y^+ + :X^- \qquad (33)$$

a) Y = H

b) Y = $-\overset{|}{\underset{|}{C}}$

In contrast to this, the defining reaction for the reduction potential of $X^{\bullet}$ (Eq. 34)

$$e^{\bullet -} + {}^{\bullet}X \rightleftarrows :X^- \qquad (34)$$

is a reaction between two open-shell systems. In the forward reaction the *radical* $X^{\bullet}$ accepts a single electron. The forward reaction involves not the shift of an electron pair betweentwo centers but the formation of an electron pair, i.e the production of a closed-shell system from two radical precursors. In view of these differences in the nature of reactions (33) and (34) a simple relation between the electron *pair* accepting ability of X in Y–X (Eq. 33 forward) and the *single* electron accepting power (reduction potential) of the radical $X^{\bullet}$ is not to be expected.

As pointed out in Sects. 2 and 3, the leaving group properties of X^- can often be changed drastically by substitution or by protonation or deprotonation (as can the redox potentials).

5 References and Notes

1. For an earlier discussion of inner-sphere electron transfer mechanisms see Ref [2]
2. Steenken S (1987) In: Fielden E M, Fowler J F, Hendry J H, Scott D (eds) Radiation Research, Proceedings 8th Internat Congr Radiation Research, Edinburgh 1987, Taylor and Francis, London, Vol. 2
3. For reviews see, e.g., a) Neta P (1976) Adv Phys Org Chem 12:2; b) Henglein A (1976) Electroanal Chem 9: 163; c) Swallow A J (1978) Progr React Kinet 9: 195; d) Steenken S (1987) J Chem Soc Faraday Trans 1, 83: 113
4. For recent reviews on electron transfer mechanisms see, e.g. a) Todres ZV (1978) Russ Chem Rev 47: 148; b) Albery W J (1980) Ann Rev Phys Chem 31: 227; c) Eberson L (1982) Adv Phys Org Chem 18: 79; d) Pross A (1985) Acc Chem Res 18: 212; e) Shaik S S (1985) Prog Phys Org Chem 15: 197; f) Eberson L, Radner F (1987) Acc Chem Res 20: 53; g) Chanon M (1987) Acc Chem Res 20: 214
5. For a discussion of single electron "shifts" in two-electron transfer processes see Refs [4d] and [4e]
6. a) Norman ROC (1970) In: Essays in Free Radical Chemistry, J Chem Soc Special Publication No. 24, London, p 117; b) Gilbert BC (1973) Electron Spin Resonance 1: 206; c) Steenken S, Davies MJ, Gilbert BC (1986) J Chem Soc Perkin Trans 2, 1003 and references therein, see also Ref [44b]
7. Koltzenburg G, Behrens G, Schulte-Frohlinde D (1982) J Am Chem Soc 104: 7311, (1983) ibid 105: 5168
8. Schulte-Frohlinde D (1983) In: Nygaard OF, Simic MG (eds) Radioprotectors and Anticarcinogens, Academic, New York, p 53
9. Solvolvsis reactions have in fact been interpreted in terms of electron transfer, cf. a) Ref. [4b–e]; b) Murdoch JR, Magnoli DE (1982) J Am Chem Soc 104: 3792; c) Chanon M (1982) Bull Soc Chim Fr 216; d) Lewis ES (1986) J Phys Chem 90: 3756
10. An excellent recent review of solvolysis reactions is to be found in Vogel P (1985) Carbocation Chemistry, Studies in Organic Chemistry, 21, Elsevier, Amsterdam
11. The validity of reaction Scheme 1 is not limited to heterolyses of covalent bonds to *carbon*. It covers also cases where both X and Y are hetero-atoms (e.g. oxidation of halides and pseudohalides by $^{\bullet}OH$, cf. Ref [40] or where X is a hetero atom and Y is a metal (cf. Ref [41])
12. For reviews see, e.g. Wardman P, Clarke ED (1985) In: Breccia A, Fowler JF (eds) New chemo and radiosensitizing drugs, Edizione Scientifiche ⟨Lo Scarabeo⟩. p 21; Wardman P (1987) In: Farhataziz, Rodgers MAJ (eds) Radiation chemistry: Principles and applications, Verlag Chemie, Weinheim, p 565
13. Minisci F, Citterio A (1980) Adv Free Radical Chem 6: 65
14. Non-"nucleophilic" radicals, e.g. alkyl radicals not substituted by a hetero atom at C_α, do not seem to react by addition to the nitro group ($k \leq 10^7\ M^{-1}s^{-1}$). There is so far no evidence that the radicals add to the benzene ring

15. Jagannadham V, Steenken S (1988) J Am Chem Soc 110: 2188
16. McMillan M, Norman ROC (1968) J Chem Soc B: 590
17. Adams GE, Willson RL (1973) J Chem Soc Faraday Trans 2 69: 719
18. Jagannadham V, Steenken S (1984) J Am Chem Soc 106: 6542
19. Steenken S, unpublished material
20. Janzen EG, Gerlock JL (1969) J Am Chem Soc 91: 3108; Sleight RB, Sutcliffe LH (1971) Trans Faraday Soc 67: 2195; Wong SK, Wan JK (1973) Can J Chem 51: 753
21. Eibenberger J, Schulte-Frohlinde D, Steenken S (1980) J Phys Chem 84: 704
22. see Lowry TH, Richardson KS (1981) Mechanism and Theory in Organic Chemistry, 2nd edn, Harper and Row, New York, p 351 ff
23. Fujita S, Steenken S (1981) J Am Chem Soc 103: 2540
24. Hazra DK, Steenken S (1983) J Am Chem Soc 105: 4380
25. Schuchmann MN, Steenken S, Wroblewski J, von Sonntag C (1984) Int J Radiat Biol 46: 225
26. Steenken S, Jagannadham V (1985) J Am Chem Soc 107: 6818
27. Jagannadham V, Steenken S (1988) J Phys Chem 92: 111
28. Steenken S, unpublished material
29. Adams GE, Willson RL (1969) Trans Faraday Soc 65: 2981; Willson RL (1970) Int J Radiat Biol 17: 349
30. For a collection of redox potentials in aqueous solution see Steenken S (1985) in: Landolt-Börnstein 13e: 147 or Wardman P (1989) J Phys Chem Ref Data 18: 1637
31. Bothe E, Behrens G, Schulte-Frohlinde D (1977) Z Naturforsch 32b: 886
32. Rabani J, Klug-Roth D, Henglein A (1974) J Phys Chem 78: 2089
33. Bothe E, Schulte-Frohlinde D (1980) Z Naturforsch 35b: 1035
34. Bothe E, Schuchmann MN, Schulte-Frohlinde D, von Sonntag C (1978) Photochem Photobiol 28: 639
35. Asmus K-D, Wigger A, Henglein A (1966) Ber Bunsenges Phys Chem 70: 862
36. Bielski BHJ, Cabelli DE, Arudi RL, Ross A (1985) J Phys Chem Ref Data 14: 1041
37. Al-Sheikley MI, Hissung A, Schuchmann H-P, Schuchmann MN, von Sonntag C, Garner A, Scholes G (1984) J Chem Soc Perkin Trans 2: 601
38. For a review see Ref [3d]
39. Schwarz HA, Dodson RW (1984) J Phys Chem 88: 3643. Kläning UK, Sehested K, Holcman J (1985) J Phys Chem 89: 760
40. Fornier de Violet P (1981) Rev Chem Intermediates 4: 121
41. Asmus K-D, Bonifacic M, Toffel P, O'Neill P, Schulte-Frohlinde D, Steenken S (1978) J Chem Soc Faraday Trans 1 74: 1820
42. Lind J, Shen T, Eriksen TE, Merenyi G (1990) J Am Chem Soc 112: 479
43. Neta P, Madhavan V, Zemel H, Fessenden RW (1977) J Am Chem Soc 99: 163
44. a) Chawla OP, Fessenden RW (1975) J Phys Chem 79: 2693; b) Davies MJ, Gilbert BC (1984) J Chem Soc Perkin Trans 2: 1809
45. O'Neill P, Steenken S, Schulte-Frohlinde D (1975) J Phys Chem 79: 2773; Walling C (1975) Acc Chem Res 8: 125; O'Neill P, Steenken S, Schulte-Frohlinde D (1977) J Phys Chem 81: 31; Sehested K, Holcman J, Hart EJ (1977) J Phys Chem 81: 1363; Sehested K, Holcman J (1979) Nukleonika 24: 941
46. Steenken S, Koltzenburg G, unpublished results
47. This number is based on the observation that in the reaction with $SO_4^{\bullet -}$ k(observed) for formation of optical density at 340 nm (where the $^{\bullet}OH$ adduct of benzonitrile absorbs) is proportional to [benzonitrile] up to the saturation limit ($\approx$ 20 mM, k(observed) = 5×10^6 s^{-1}), as shown by 248 nm laser experiments
48. Hasegawa K, Neta P (1978) J Phys Chem 82: 854
49. Jayson GG, Parsons BJ, Swallow AJ (1973) J Chem Soc Faraday Trans 1 9: 1597
50. Littler JS (1970) In: Essays in free radical chemistry, J Chem Soc Special Publication No. 24. London, p 383

Received: May 1995

Antenna Structure and Energy Transfer in Higher Plant Photosystems

Robert C. Jennings[1], **Roberto Bassi**[2], **and Giuseppe Zucchelli**[1]

[1] Centro CNR Biologia Cellulare e Molecolare delle Piante, Dipartimento di Biologia, Universita' degli Studi di Milano, via Celoria 26, 20133 Milano, Italy.
[2] Dipartimento di Biotecnologie Vegetali, Universita' di Verona, Strada Le Grazie 5, 37100 Verona, Italy.

Table of Contents

The light harvesting antenna of higher plant photosystems contains a large number of chlorophyll and carotenoid pigments which are bound to a number of different polypeptides. Absorbed light energy is transferred with high efficiency from the antenna to reaction centres where primary photochemistry occurs. In this review, a detailed examination of the biochemical characteristics of these pigment-proteins is undertaken and a topological structural model for photosystem II is presented. Energy transfer between chlorophylls is discussed in terms of dipole coupling between spectral forms which are homogeneously and inhomogeneously broadened to different extents. Reasonable agreement is found between calculated and measured pairwise transfer rates which are in the order of 1 ps. Energy transfer between chlorophylls and carotenoids is discussed in terms of both dipole and exchange coupling. The energy funnel organisation of spectral forms seems not to be important for higher plant photosystems. Excited state spectral equilibration is expected to be considerably faster than antenna-reaction centre equilibration. Exciton dynamics are discussed in terms of trap and diffusion limited models. Slow reaction centre trapping and the absence of strong energy "funnelling" in photosystem II may be important for the down regulation of excited state levels which protects against photoinhibition.

List of Abbreviations

A, Absorption; chl, chlorophyll; car, carotenoid; EET, excitonic energy transfer; EF, exoplasmic fracture face; EM, electron microscopy; FWHM, full width at half maximum; IEF, Isoelectric Focusing, LD, linear dichroism; LHC, light harvesting complex; PAGE, polyacrylamide gel electophoresis; PF, protoplasmic fracture face; PS, photosystem; RC, reaction centre; SDS, sodium dodecyl sulphate; SSTT, single step transfer time.

1 Introduction

Higher plant thylakoids contain a number of pigment binding proteins which are organized into two photosystems. The major light harvesting pigments are chlorophyll molecules. The size of this pigment array is somewhat variable with the growth conditions but typical values of around 200 and 250 chlorophyll molecules respectively for Photosystem I and II can be obtained from many measurements [1, 2] (see [3] for a review). Each photosystem can be envisaged as having two parts: (1) a core complex built up of *chla* binding proteins, where both light harvesting and electron transport functions are performed, surrounded by (2) the outer antenna which contains several *chla/b* binding proteins. The core complex subunits are chloroplast encoded while the LHC ones are coded by the nuclear genome, translated in the cytoplasm and imported into the chloroplast as precursors. These are inserted into the thylakoid membrane in a process which includes the folding of the peptides, the binding of pigments and the cleavage of the proteins to their final size [4]. The light energy absorbed by the antenna pigments is transferred to special reaction center pigments, located in the core complex and indicated as P700 and P680 respectively in the case of PSI and PSII, where the primary charge separation/electron transport reactions occur.

The polypeptide moieties of the pigment-proteins function in the orientation and spacing of their bound pigments so that the energy absorbed by any of these pigment-proteins is transferred efficiently to the reaction center. A complementary function for these polypeptides has been recently recognized in regulatory mechanisms involving modifications in the structure of both the protein and pigment moieties.

It is intended in the present review to critically summarize current knowledge concerning structure and function of the pigment-protein complexes of higher plant photosystems.

2 Photosystem II: Composition and Organization of Chlorophyll-Proteins

Photosystem II is located in the stacked membranes of granal chloroplasts where it forms the large EFs (11.7×15.5 nm) freeze-fracture particles with the core complex while the outer LHCII is arranged in the complementary fracture face (PFs) to form 9.0×10.3 nm particles [5, 6]. PSII is composed of chl*a* binding core complex and several surrounding *chla/b* proteins which constitute the outer antenna. The whole PSII complex can be prepared as stacked membranes free of other thylakoid complexes [7]. When solubilized, the complex splits into the core complex and the outer antenna components that can be separated by sucrose gradient ultracentrifugation or PAGE [8–10]. The PSII membranes and the isolated core complex can catalyze electron transport from water to quinone analogues or other electron acceptors (for a review see [11]).

2.1 The Photosystem II Core Complex

The core complex binds the electron transport cofactors Mn^{++}, P680, phaeophytin, Qa in addition to 50–55 antenna chl*a* molecules, as well as some carotenoids (Table 1). The two sets of cofactors are bound to distinct polypeptides with electron transport components being bound to the D1 and D2 polypeptides while the antenna pigments are located on the two homologous CP43 and CP47 proteins. Additional subunits of the core complex include the two cytochrome b559 subunits and the four oxygen evolving enhancer (O.E.E.) polypeptides which are nuclear encoded. Chlorophyll binding proteins in PSII thus include the D1 and D2 polypeptides which are encoded respectively by the *psbA* and *psbD* genes. Both proteins are composed of 353 residues in most species although their apparent molecular weight in SDS-PAGE is respectively of 32 and 34 kDa.

In the light of the unequivocal location of PSII reaction centre in the D1-D2-Cyt b559 complex [12] both CP43 and CP47 must be considered as a part of the light harvesting system. They bind chl*a* and carotenoids (Table 1) and are encoded by the *psbB* and *psbC* genes which are located in the chloroplast DNA close to the *psbA* and *psbD* genes coding for D1 and D2 proteins. The deduced protein sequences are well conserved, with homology being 94 and 95% between higher plant proteins and 72 or 77% with the cyanobacteria proteins in the case of CP47 (508 residues) and CP43 (461 residues) respectively. Hydrophobicity plots suggested seven [13] or six [14] transmembrane helices for CP47 while five [15] or six [16] are predicted for CP43. However, in the light of the significant homology between the two proteins, a common structure with six transmembrane helices can be hypothesised. The two proteins are thought to bind 20–25 chl*a* molecules each [11, 17], though lower values have been suggested [18] on the basis of functional measurements in developing plant material and chl*b*-lacking mutants or biochemical measurements [19]. The lower values should be viewed with caution in the light of the recent finding that intermittent light grown plants show changes not only in the subunit stoichiometry but also in the number of chl molecules per polypeptide [20]. A more complete study on the pigment binding to PSII core subunits is certainly needed.

Table 1. Pigment-proteins of the PSII antenna system from higher plants

complex	gene(s)	coding site	% of total PSII chl	chl *a*/*b* ratio	car
D1	psbA	cp	1–2	–	ND
D2	psbD	cp	1–2	–	ND
CP47	psbB	cp	8–10	–	9.9
CP43	psbC	cp	8–10	–	17.0
LHCII	*Lhcb*1	nu	60–65	1.45	30.7
	*Lhcb*2	nu			
	*Lhcb*3	nu			
CP29	*Lhcb*4	nu	6	2.8	44.9
CP26	*Lhcb*5	nu	6	2.2	38.5
CP24	*Lhcb*6	nu	3	1.6	59.7

car, carorenoid: values are in moles per 100 moles of chl*a*. cp, chloroplast; nu, nuclear

2.2 The Photosystem II Outer Antenna

The major LHCII complex was the first chl-protein described [21]. It constitutes about one third of the total thylakoid protein and binds half of the total thylakoid chlorophyll. This protein is mostly, if not completely, present in oligomeric form in the membranes [8, 9]. The pigment complement of LHCII has been estimated biochemically to be between 12–15 chl molecules per polypeptide monomer with a chl*a*/*b* ratio of 1.45 [10, 22]. Electron crystallography suggests the presence of 14–15 chl per polypeptide. The xanthophylls neoxanthin and lutein have been reported to be present in different ratios [9, 23]. These discrepancies probably are due to the strong heterogeneity that can be found in this complex. Not only several polypeptides with very similar characteristics can be resolved from the LHCII complex by denaturing electrophoresis [24–26] but also several LHCII subpopulations can be resolved by IEF [27, 28]. This is consistent with the large number of highly homologous genes which have been found in several species [29, 30] and which fall into three types known as *Lhcb*, *Lhcb*2 and *Lhcb*3 [31]. Due to the strong immunological cross reactivity, the very high homology and the N- terminal block of all but the *Lhcb*3 gene products, it is difficult to establish the polypeptide-gene correspondence for LHCII. Moreover at least two additional sources of heterogeneity have been proposed due to post- translational modification [32, 33] and cleavage at different maturation sites [34]. The problem of the source of heterogeneity has been addressed by the two complementary approaches of direct protein sequencing [35] and immunoblot with antibodies raised against divergent peptides deduced from gene sequences [26], giving rise to the suggestion that each polypeptide is a distinct gene product. In maize three of the six major LHCII polypeptides were shown to be products of *Lhcb*1 genes and one each of *Lhcb*2 and *Lhcb*3. There might be additional *Lhcb* gene types coding for the polypeptides belonging to this complex.

The two methods which have proven to be most effective in providing information on the structure of the antenna proteins have been sequence analysis and electron crystallography. Following the cloning of the first *Lhcb*1 and *Lhca*1 genes it was recognized that a common structure could be predicted for all *Lhc* polypeptides, on the basis of hydrophobicity plots and other secondary structure prediction algorithms. This structure has three transmembrane alpha helices with the N-terminal on the stromal side and the C-terminus on the opposite surface. The first and the third helices are longer than needed for membrane spanning and therefore presumably extend into the stromal side more than the helix II [36–38]. The first and third helices are highly conserved and homologous to each other thus suggesting that the 1hc family derives from a duplication event of an ancestral gene coding for a two membrane spanning polypeptide. This was recently supported by the cloning of the *psbS* gene, homologous to the 1hcs, but coding for a four-times membrane spanning polypeptide [39, 40]. Two highly conserved, hook like, secondary structures are also present before helices I and III. A careful analysis of the deduced sequences of all 1hc gene types has been recently carried out [41] showing that the predicted structure is very similar in all cases while differences can be expected in the N-terminus, in the loop on the stromal side between helices I and II, which is expected to be reduced in length in the case of *Lhcb*1, 2, 3 and 5 gene products. The structure of the LHCII polypeptide has been proposed to reversibly change upon phosphorylation in the N-terminal threonine residue during the process of State transitions [42]. In this process, the negative charges added would move the N-terminal to a positively charged pocket formed by the stromal extensions of helices I and III [43]. This structural change may be the basis of the phosphorylation- induced dissociation of LHCII trimers from the CP29-CP24 complex and for migration of the monomers to stroma membranes [44–49].

The predictions from sequence analysis have been confirmed by electron crystallography which revealed the three dimensional LHCII structure at 6 Å and at 3.4 Å resolution in projection [50, 51] (Fig. 1). The two homologous helices are shown to be tilted with respect to membrane plane and cross each other in a X shaped structure. The third helix is separate and perpendicular to the membrane plane. At this resolution it is proposed that 14 or 15 chl molecules can be detected as distinct regions of electron density not connected to the polypeptide chain. The chl rings are arranged on two levels, each located at the position of the hydrophobic hydrocarbon chains of the lipid bilayer, and are oriented at about 10° with respect to the membrane normal. The center to center distances between nearest neighbors on each level are in the 9–14 Å range. When observed in the membrane plane the porphyrins of the upper level are arranged in a ring of eight so that each has two neighbors. The ring plane is inclined so that three out of eight chls are closer (13–14 Å) than the others to three of the 6–7 chls of the lower level. Three of the above described monomers are organized into a trimeric structure within the crystal lattice which probably corresponds to the oligomeric form characteristic of the major LHCII when isolated [8, 22] by mild PAGE or sucrose gradient ultracentrifugation. By contrast the minor chl-proteins have a lower aggregation state [9, 10] which is also obtained after dissociation of LHCII [22]. It is most likely that the major LHCII has a trimeric organization while the minor chl-proteins are monomers.

*Lhcb*1 was the first Lhc cDNA identified and sequenced [52, 53]. Since then sequences have been made available from 20 species (for a review see [54]) *Lhcb*1 is present in many copies in the genome ranging from 5 in *Arabidopsis* [55] to 16 in *Petunia* [29]. The mature protein consist of 232 amino acids starting with an acetylated Arginine residue [56]. However small differences exist between species and between different *Lhcb* genes within the same specie thus yielding multiple *Lhcb*1 polypeptides [26]. The protein is reversibly phosphorylated at a threonine residue close to N-terminal [33, 57].

The *Lhcb*2 gene codes for a polypeptide slightly smaller with respect to the one *Lhcb*1 codes for. However *Lhcb*2 and *Lhcb*1 are highly homologous (up to 90%). Consistently, the predicted structure is almost the same, apart from a

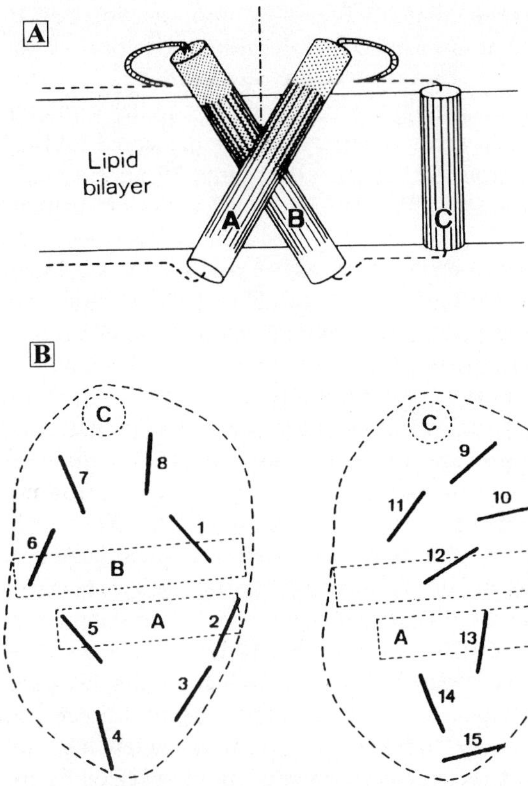

Fig. 1. The structure of the LHCII monomer as derived from electron crystallography [51]. **A** proposed topography of the polypeptide in the photosynthetic membrane. Letters *A*, *B* and *C* indicate the three hydrophobic α-helices spanning the membrane. Chlorophyll molecules are arranged into two rings roughly parallel to the membrane plane. **B** Approximate position of the chlorophyll in the upper level (*left*) and lower level (*right*) on the membrane plane. *Dashed lines* outline α-helices A, B and C. Chlorophyll molecules are oriented perpendicular to the membrane plane and are thus represented as *black bars*. Chlorophylls numbered as *6*, *7* and *8* are closer to those belonging to the lower layer than the other pigment molecules

slightly reduced N-terminal stromal extension [41]. The actual length is about 228 amino acids, although N-terminal blocking prevents precise identification of the N-terminus. *Lhcb*2 genes have been identified from many species (for a list see [31]) and are present in 1-2 copies per haploid genome and carry one intron. Jansson et al. [35] have determined that the *Lhcb*2 polypeptide is the lower component of the doublet resolved by Laemli SDS-PAGE, which is less abundant in the thylakoids with respect to the upper (*Lhcb*1) gene product. In maize, out of the major six LHCII polypeptides, only one was identified as *Lhcb*2 product versus three *Lhcb*1. This polypeptide is the most rapidly phosphorylated during State transitions [46, 58, 59] and the most readily depleted from grana stacks upon phosphorylation [28]. These data and the finding that the relative abundance of *Lhcb*2 is increased upon acclimation to low light regimes [27, 60], suggest it is enriched in the outermost part of the antenna.

Although less homologous than *Lhcb*2, *Lhcb*3 shares 78 to 80% of the residues with *Lhcb*1 and *Lhcb*2 thus yielding a very similar predicted folding. The major difference is in the deletion of the first N-terminal 12 amino acids, which includes the phosphorylation site [55, 61–63]. *Lhcb*3 is not part of the mobile LHCII pool [48].

The three minor chl*a*/*b* proteins of PSII, CP29, CP26 and CP24, are coded by the nuclear genome and are thought to be located in between the core complex and the major LHCII (see Sect. 2.3). It is not yet clear whether they partition with EFs or PFs particles in freeze fractured thylakoids. They have an higher chl*a*/*b* ratio with respect to LHCII and bind only about 15% of the total PSII chl vs 63% of LHCII [10, 41]. Quantitative analysis has shown that these proteins bind more lipids with respect to both the major LHCII and the PSII core complex [64]. These pigment-proteins have been recently shown to bind together more than 80% of the xanthophyll violaxanthin [23], the precursor of zeaxanthin, which has been involved in the process of non-photochemical quenching of excess excitation energy [65]. This result strongly suggests a role for these proteins in the regulation of energy transfer to PSII core complex.

The pigment-protein complex CP29 was the first minor complex to be distinguished from LHCII [66]. Its apparent mass is slightly higher (31 kDa) than that of the major LHCII polypeptides both in denaturing and green gels [8, 67]. The protein is N-terminally blocked but partial sequencing was obtained from spinach, maize and tomato [68] showing that it is coded by the *Lhcb*4 gene which was recently isolated and sequenced [61]. This is the largest Lhc gene, coding for a mature protein of approximately 257 amino acids, and is extremely well conserved between maize and barley not only in the mature protein but also in the pre-sequence region. The larger size is due to a 42 residues long insertion, which is not coded in other Lhc genes, located just before the first transmembrane helix. This complex has a chl*a*/*b* ratio of about 2.2–2.8 [9, 10, 69] and contains lutein, violaxanthin and neoxanthin as additional pigments [9, 23]. Values of four [70], eight [10] and ten [68] chl molecules per polypeptide have been reported. The lower values may be due to the loss of pigments during isolation. The Qy absorption maximum is at 677–678 nm [68, 69, 71] while a characteristic small peak is visible at 641 nm (see Sect. 4). Fluorescence emission is characterized by a major peak with a maximum at 681.5 nm [72]. CP29 is almost certainly the protein that has been shown to bind Ca^{++} in spinach [73] and is related to a polypeptide with slightly higher apparent molecular weight which accumulates in maize following chilling in the light as shown by antigenic cross reactivity [74].

The CP26 pigment-protein complex has been described in maize and spinach as having an intermediate chl*b* content between CP29 and LHCII [8, 75, 76]. Its pigment complement includes violaxanthin, lutein and neoxanthin as well as chl*a* and chl*b* in a 2.2 ratio [10]. Lower (1.8) and higher (2.7) *a*/*b* ratio values have also been reported [41, 70]. Between 9 to 11 chl molecules per polypeptide have been determined [9, 10]. In urea gels two closely migrating apoproteins are resolved, in similar amounts, with an apparent mass of 28 and 29 kDa [8]. Both polypeptides are N-terminally blocked and therefore the actual molecular weight and the maturation site are not known. Antibodies against oligopeptides obtained from the *Lhcb5* gene recognize CP26 apoproteins [77] thus implying that this is the coding gene. This has been confirmed by direct protein sequencing. *Lhcb5* genes have been sequenced from tomato, barley, pine and *Arabidopsis* [36, 78, 79]. It has been reported that the product of *Lhcb5* gene also codes for a CP29 component [78], however, this was not confirmed by analysis of the purified complex [31]. This result may be attributed to the comigration of CP26 and CP29 in tomato. CP26 has been reported to be the main Cu^{++} binding protein in thylakoids [80].

CP24 is the product of the *Lhcb6* gene as shown by N-terminal sequencing [81]. *Lhcb6* genes have been sequenced from tomato, spinach and maize [55, 82] coding for a 210 amino acid protein. The carotenoid content is qualitatively similar to that of CP29 and CP26 but for the absence of neoxanthin, while very divergent values have been reported for the chl*a*/*b* ratio (1.6–0.8) and the chl to polypeptide ratio (5–13) [9, 10] due presumably to the preferential loss of chl*a* during PAGE. Its red absorption peak is at 675.5 nm and the fluorescence emission at 681.5 nm [71, 72]. The characteristics of the protein with regard to molecular mass, pigment composition and immunological cross reactions are very similar to the LHCI-680 component *Lhca2* thus leading to the suggestion, in earlier reports, that they were the same protein [8]. This hypothesis was later disproved by using monoclonal antibodies [24] and N-terminal sequencing [61]. However a polypeptide very similar to CP24 is present in LHCI as recently found in *Chlamydomonas reinhardtii* [83].

Following several earlier reports [84, 85] the finding of a low molecular mass chl*a*/*b* protein, LHCIIe, has been reported [9] which is greatly enriched in xanthophylls. To date, other labs have not been able to confirm this interesting report. This possibility exists that the above reports could be due to the presence of a new class of proteins transiently synthesized during light exposure of dark grown plants [86] or after photoinhibition [87] but also present at low levels in fully greened leaves [20]. N-terminal sequencing of LHCIIe is needed in order to verify this hypothesis.

The gene coding for a 22 kDa protein, previously reported to be important in PSII assembly [88], has been sequenced [39, 40] and an homology with *Lhcb* genes was recognised although a four, rather than a three helix structure is hypothesized on the basis of hydrophobicity plots. Pigment binding to this polypeptide has not been demonstrated.

2.3 The Organisation of PSII Chlorophyll-Proteins: a Model for the Structure of the Antenna System

The supramolecular organization of PSII within the photosynthetic membrane is not yet clear. However, in the past few years, some information has been obtained about the aggregation state of the individual chl-proteins, their association into heterodimers and their topological organization, which can be used in defining a tentative model of PSII antenna system. The core complex is composed of up to 18 polypeptides. However many are small or are soluble proteins located on the lumenal surface of the membrane. The general organisation of the core is expected to be determined by the four major subunits D1, D2, CP43 and CP47 which also bind both the electron transport and antenna cofactors and are present in the same molar ratio. Recently evidence has accumulated which suggests that the PSII core is dimeric. Electron microscopy and image analysis of two-dimensional crystals [89, 90] have shown that the mass of the complex is around 600–700 kDa. Since the sum of the molecular weights of the component polypeptides is about 250 kDa, this result strongly suggests that the core complex is present as a dimer in PSII. This is in agreement with other ultrastructural work [91–93] but contrasts with the results of de Vitry et al. [17, 94, 95] who found mostly monomeric PSII in *Chlamydomonas reinhardtii.* In a very recent study [96] a three dimensional reconstruction of PSII EFs particles has been described. The authors suggest that the structure resolved corresponds to a reaction center core complex surrounded by external antenna proteins, on the basis of the fact that the molecular weight estimation fits with the expected mass of a reaction center surrounded by external antenna proteins [10]. However it was shown in an earlier report that the shape of the EFs particles is only slightly affected in the Chlorina f2 mutant lacking the antenna proteins [97] which are expected to constitute half of the mass of a PSII-LHCII complex [10]. Moreover it was shown that the detergent treatment used in the particle array preparation produces lateral segregation of the LHCII to other membrane regions where it forms distinct patches of particles [98]. Thus it is unlikely that LHCII is part of the structure described. Biochemical analysis by Deriphat-PAGE supports the dimeric hypothesis for the PSII in grana membranes while the small PSII pool in the stroma membranes seems to be monomeric [99].

On the basis of quantitative polypeptide and pigment analysis the following stoichiometry has been suggested for PSII antenna (PSII core/CP29/CP26/CP24/LHCII):1 : 1 : 1 : 1 : 12 [9]; 1 : 1.5 : 1.5 : 1.5 : 12 [10]; 1 : 2 : 2 : 2 : 14; [100, 101]. These differences may be in part due to differences in the species used and in growth conditions. The values can be used to calculate the PSII antenna size. Values which come out for chls per photosystem from the stoichiometries are respectively 246, 255 and 294 assuming the following chl/polypeptide ratios: LHCII, 14; PSII core, 53; CP29, 8; CP26, 9; CP24, 6. The first two values are in close agreement with experimental determination [1–3].

When PSII is solubilized in mild conditions, besides the above described trimers of LHCII, complexes containing two or more different pigment-proteins can be isolated. This provides information on the topological organization of PSII. The most interesting of these contains CP29, CP24 and LHCII in 1 : 1 : 3 molar ratio [10, 63]. This complex splits into a CP29-CP24 and a LHCII moiety upon phosphorylation [49, 63]. A second group of complexes contains the PSII core and one or more of the surrounding antennas: CP26 and CP29 have been shown to be present in the preparation depleted of both LHCII and CP24 [70, 102]. This approach was further developed by Tidu [103] who isolated four different PSII core-LHC complexes with increasing size and LHC polypeptide complexity by electroendoosmotic electrophoresis. Their composition is consistent with a model in which LHCII is mostly, if not exclusively, connected to the core through one or more of the minor antenna complexes. This view is in agreement with the finding that antennas are added in discrete steps during development [101] whose size is consistent with the simultaneous connection of both a minor complex and an LHCII trimer.

The above described data have been summarized by different laboratories into structural models. In Fig. 2 the most recent proposals are shown. They all agree on the point that the minor pigment-proteins should be located in a pericentral position between the core complex and the major LHCII. Major differences are encountered in the monomeric (B) or dimeric (A, C, D) organization of the core, the stoichiometry between the core complex and LHC proteins (model in panel D uses the same stoichiometry as the one in panel A) and, finally, the possibility that CP29 and CP24 belong to the same subcomplex (A, C), as previously suggested [9, 10, 63], or are independently connected to the core complex (B, D).

3 Photosystem I: Composition and Organization of Chlorophyll-Proteins

PSI is a multisubunit complex which is located in the unstacked, stroma exposed membranes where it forms the large PFu particles (10.3 × 12.5 nm) visible by freeze fracture E.M. [104]. It is composed of a core complex (PSI core) and a light harvesting component (LHCI). A particle containing both these components can be obtained by solubilization with anionic detergents and sucrose gradient ultracentrifugation with a chl/P700 ratio of about 200 and a chl*a*/*b* ratio of 6–6.5 [75, 105, 106]. This preparation contains at least 15 polypeptides. Similar preparations, although in smaller amounts, can be obtained by SDS-PAGE or Deriphat-PAGE [107, 108]. Several pigment-proteins are contained in this complex which are described below.

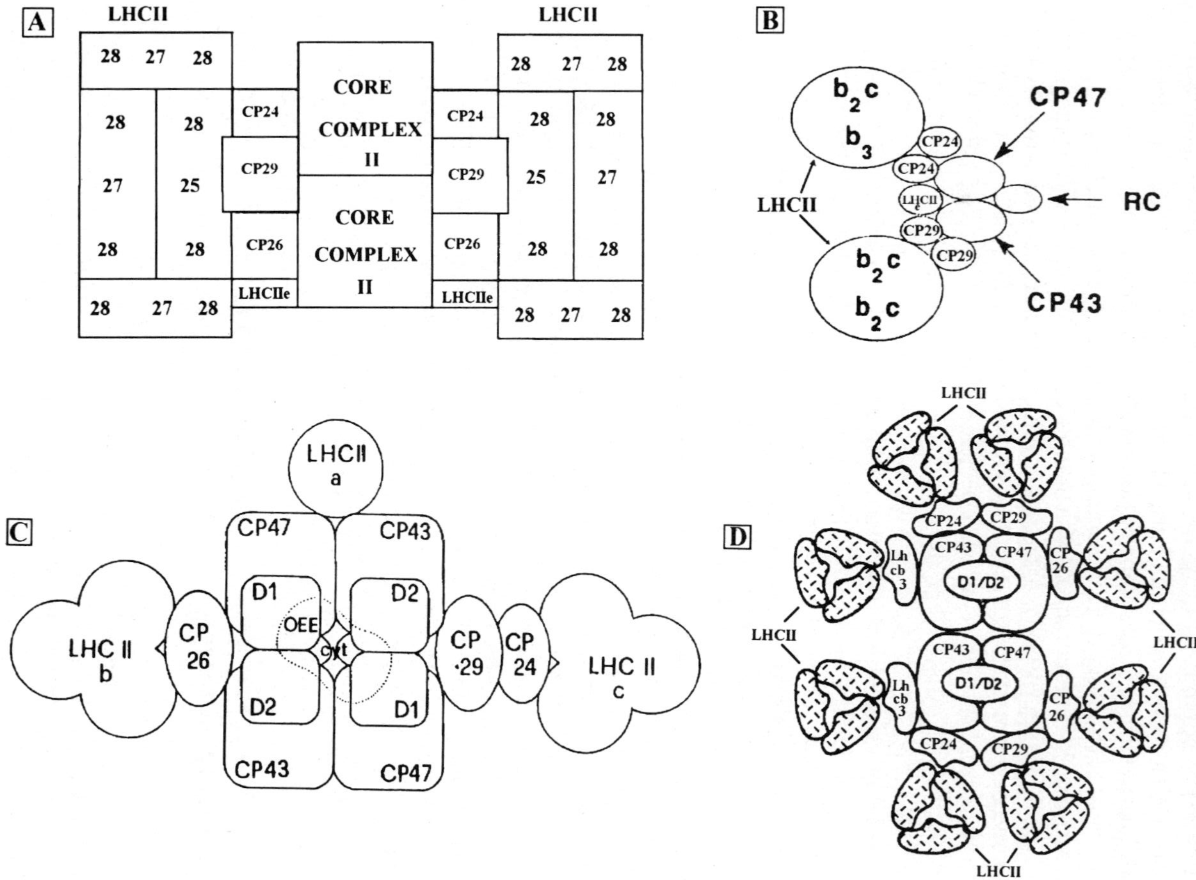

Fig. 2. Models for the organisation of PSII in the membrane plane. Model in (**A**) is from Peter and Thornber [9]. Model in (**B**) is from Harrison e Melis [100]. b2c and b3 indicates two kinds of LHCII trimers with and without the *Lhcb*3 gene product. Model in (**C**) is from Dainese et al. (1992) [251] LHCII *a*, *b* and *c* indicates respectively the unphosphorylateable-non mobile, the phosphorylateable-non mobile and the phosphorylatable-mobile pools of LHCII. Model in (**D**) is from Jansson (1993) [252]. For the sake of simplicity the same nomenclature has been used, where possible, to indicate the complexes

3.1 Photosystem I Core Complex

This is a chl*a* binding complex which has an apparent size of 250 kDa in non-denaturing PAGE [9, 75, 107]. This particle, which binds P-700, can photoreduce NADP in the presence of ferredoxin and ferredoxin-NADP reductase [109]. All pigments are bound to the two major protein subunits which are the products of psaA and psaB chloroplast genes with P700 being at the interface between the two [110]. The two homologous subunits together bind about 90 chl*a* molecules [75].

3.2 Light Harvesting Complex I

In higher plants, LHCI splits into two moieties, called LHCI-680 and LHCI-730 according to their fluorescence emission maxima measured at 77 K, when detached from PSI core [111]. Most authors agree that LHCI polypeptides are the product of four *Lhca* genes (discussed below). The first report of LHCI isolated as a high molecular weight complex, is from *Chlamydomonas reinhardtii* [112]. In this organism the complex appears to be somehow different with respect to the case of higher plants in that the fluorescence emission of the whole complex is at 705 rather than 735 nm [112]. Moreover, the number of LHCI polypeptides is higher (7–10 vs 4–5 in higher plants) [83]. There are no rigorous measurements for the pigment-protein stoichiometry in both LHCI-730 and LHCI-680 due to the loss of pigments which occurs when dissociating individual proteins from the LHCI complex. Indirect evidence suggests that each LHCI polypeptide may bind 8–10 chl molecules [83]. The composition of higher plants LHCI is summarized in Table 2.

N-terminal sequencing of the two polypeptides present in spinach and barley LHCI-730 allowed their identification with *Lhca*1 and *Lhca*4 gene products

Table 2. Pigment-protein composition of PSI antenna system from higher plants

complex	gene(s)	coding site	% of total PSI chl	chl *a*/*b* ratio
PSI core	psaA	cp		
	psaB	cp	45	—
LHCI-730	*Lhca*1	nu		
	*Lhca*4	nu	35	3.0
LHCI-680	*Lhca*2	nu		
	*Lhca*3	nu	20	3.0
psaF	psaF	nu	low	> 6

cp, chloroplast; nu, nuclear

[113, 114]. *Lhca*1 codes for a 200 amino acids polypeptides in *Arabidopsis*, tomato, tobacco and Scots pine [35, 115–117] which has a high degree of homology with the *Lhcb*4 gene coding for the PSII antenna CP29. *Lhca*4 gene codes for 199–200 amino acid long mature proteins in tomato, *Arabidopsis* and Scots pine [35, 62, 118]. The two proteins form a complex binding chl*a*, chl*b*, violaxanthin, lutein and traces of neoxanthin and beta carotene [119]. The chl*a*/*b* ratio has been reported in the 2.3 to 3.0 range and the complex has been shown to bind 66 chl*a* and 22 chl*b* molecules [75, 120].

The LHCI-680 complex has been obtained by sucrose gradient ultracentrifugation [75, 111] or non-denaturing PAGE [107, 121] and contains two polypeptides later identified as *Lhca*2 and *Lhca*3 gene products [113, 114]. The evidence that these two polypeptide form a complex is not very strong because the two methods used are not expected to resolve comigrating polypeptides with similar molecular masses, as is probably the case for *Lhca*2 and *Lhcb*3 gene products. It is therefore probable that LHCI-680 represents two independent chlorophyll proteins as also suggested by the electrophoretic data of Knoetzel et al. [114]. *Lhca*2 sequences have been obtained from petunia [122], tomato [123] and Scots pine [35] showing that they code for a 211 amino acid-mature protein. *Lhca*3 cDNA clones have been isolated from tomato [124] and pine [35] and code for a preprotein of 286 residues whose mature size (probably 232 amino acids) is not known with certainity due to N-terminal blocking. Very different values for chl*a*/*b* ratio have been suggested, ranging from 1.4 to 3.0 [75, 107, 120]. The lower values are probably due to the fact that some pigment proteins like LHCI-680 and CP24 are particularly prone to loss of chl*a* when subjected to PAGE. The number of pigments bound to LHCI-680 in a PSI unit is about 30–35 [75].

Although most authors believe that LHCI polypeptides are coded by a multigene family whose components have been described above, it has recently been reported that the product of psaF, an unrelated gene, previously indicated as plastocyanin binding protein is indeed a chl*a* binding subunit [120] which is part of LHCI. The deduced sequence of the psaF [125] subunit is not related to that of the other antenna polypeptides.

4 Spectral Forms

It is useful to define the term spectral form as it is used in this review. Whilst a rigorous definition might refer to a homogeneously (vibrationally) broadened site electronic transition, we prefer a less limiting definition which allows for a site distribution of closely lying transitions (site inhomogeneous broadening) within a single apparently structureless absorption band. Thus the band width (Γ) of a spectral form may be defined [126], in the case of gaussian bands, by

$$\Gamma^2 = \Gamma^2_{\mathrm{hom}} + \Gamma^2_{\mathrm{inh}} \tag{1}$$

In physical terms the site inhomogeneous component may be conceived of either as slightly different protein binding sites for chromophores or as a distribution of protein conformational substates [127] at any one site. In operational terms the spectral forms are represented by the sub-bands of a gaussian or lorentzian decomposition analysis of absorption or fluorescence spectra.

Whilst chlorophyll absorbs light over the entire visible spectrum, relaxation to the lowest lying excited singlet state is extremely rapid. Thus singlet energy transfer occurs via the S_0–S_1 (Q_y) transition. Unless otherwise stated all comments made here on chl spectroscopy concern this transition.

The absorption of chl*a* bound to polypeptides in the pigment-protein complexes of PSI and PSII (see section 2 and 3) is spectrally broadened by 10–50% with respect to chl*a* monomers in organic solvents at room temperature. The broadening is generally thought to be due to a coarse-grained site inhomogeneous broadening associated with a number of strongly overlapping spectral forms, each with a distinct absorption maximum [128–133] in the 660–720 nm range. This interpretation of chl-protein complex absorption in terms of spectral forms is based largely on derivative spectroscopy and sub-bands analysis techniques in terms of gaussian or lorentzian components. On the other hand Leupold et al. [134] observed non-linear absorption changes, in pump-probe measurements at high monochromatic pump intensities, over a very broad wavelength band in chloroplasts. The results were interpreted in terms of the entire $Q_y(0, 0)$ band of chl in chloroplasts being essentially homogeneously (vibrationally) broadened with no spectral forms. However narrow (< 1–$3\ cm^{-1}$) zero phonon holes have been burned in both absorption and emission spectra over essentially the entire $Q_y(0, 0)$ band at low temperatures (1.5 K–4 K) in chl antenna systems [135–141]. These data provide unequivocable evidence that the chl*a* band in chl-protein complexes is inhomogeneously broadened. The very broad non-linear absorption changes described by Leupold et al. [134] may be explained by rapid energy transfer over the entire antenna during the nanosecond pulse-probe analysis, even thought excitation was red-shifted by 20 nm with respect to the absorption peak. Efficient "uphill" energy transfer has been demonstrated in PSII antenna [142, 143].

The relation between the narrow spectral holes burned in plant antenna systems and the spectral forms is not very clear at the moment. The FWHM of the spectral forms is in the 10–17 nm (200–300 cm^{-1}) range at room temperature [71, 144, 145]. It is possible to calculate a lower limit of the bandwidth due to homogeneous broadening for the C670 PSI RC complex using the Huang-Rhys factor for the electron-phonon coupling ($S \approx 0.8$) and the mean phonon frequency for antenna-chl*a* suggested by Hayes et al. [126]. With the expressions of Hayes et al. [126, eq. 28, 29] a value of 200 cm^{-1} (about 10 nm in the Q_y chl spectral region) comes out for $T = 300$ K, thus indicating the importance of this factor. These authors have suggested for this particle that the site inhomogeneous broadening is about 200 cm^{-1}. However this value seems to refer to the coarse-grained inhomogeneous broadening associated with the spectral forms and not an inhomogeneous broadening of each individual spectral form. For LHCII the 678 nm and 670 nm forms display considerable band narrowing upon lowering the temperature, while the 660 nm band is not very temperature sensitive [132]. This suggests that there may be rather different homogeneous and inhomogeneous contributions to spectral broadening of the different spectral forms.

The physico-chemical interactions giving rise to the spectral forms in chl-antenna complexes are not known. Gudowska-Nowak et al. [146] have shown from INDO method calculations that large spectral shifts can be expected due to conformational variations of the porphyrin skeleton imposed by the apoprotein. It has also been suggested that the long wavelength forms, such as that absorbing near 684 nm, may result from chl-chl interactions, on the basis of in vitro studies on chl dimers [147]. Recent results indicate that in all six complexes of PSII antenna the same spectral forms seem to be present at approximately similar relative levels, as judged by gaussian band analysis [71]. Whilst this is readily understandable for the four highly homologous lhcb complexes of the outer antenna, it is less so for the core complexes which are structurally dissimilar (Sect. 2). These observations may prove useful in suggesting possible interactions for the spectral forms when more is known on the chl binding sites. In this context it should be mentioned that a precise hypothesis has been formulated by Brunisholz and Zuber [148] to explain bacteriochlorophyll spectral forms in antenna complexes of purple bacteria in terms of protein-bacteriochlorophyll interactions. In particular it was suggested that interactions with nearby aromatic amino acids may be important in determining the spectroscopic properties of red shifted core antenna complexes. Recently site-directed substitution of specific amino acids [149] and also specific proteolysis of pigment-protein complexes [150] have provided strong support for this concept.

It has been known for many years, mainly on the basis of linear dichroism studies, that there are significant differences in orientation of the Q_y (0, 0) transition across this absorption band in thylakoids. From the earlier literature [see e.g. 151] the following general conclusions could be drawn. The long wavelength chl*a* forms ($\lambda > 676$ nm) have their Q_y transition dipoles lying fairly close to the membrane plane, while the shorter wavelength chl*a* forms are closer to 35° out of the membrane plane. Most chl*b* is thought to be oriented at angles greater than 35° out of the membrane plane. More recently this picture has been confirmed and somewhat refined for various antenna complexes of both photosystems [131, 152–156]. Attempts to extract the orientation factors associated with the spectral forms from LD measurements have relied on calculation of the dichroic ratio (LD/A) or related parameters. This analysis cannot however yield exact information as the LD signal is very sensitive to the degree of sample orientation, which can be extremely variable. In addition the measured absorption and orientation at each wavelength is a linear combination of a number of overlapping spectral bands. This second problem may at least in part be obviated by performing spectral decomposition of LD spectra.

Little is known about the fluorescence of the chl*a* spectral forms. It was recently suggested, on the basis of gaussian curve analysis combined with band calculations, that each of the spectral forms of PSII antenna has a separate emission, with Stokes shifts between 2 nm and 3 nm [133]. These values are much smaller than those for chl*a* in non-polar solvents (6–8 nm). This is due to the narrow band widths of the spectral forms, as the shift is determined by the absorption band width for thermally relaxed excited states [157]. The fluorescence rate constants are expected to be rather similar for the different forms as their gaussian band widths are similar [71]. It is thought that the fluorescence yields are also probably rather similar as the emission of the spectral forms is closely approximated by a Boltzmann distribution at room temperature for both LHCII and total PSII antenna [71, 133].

The presence of different chl*b* spectral forms has not been unequivocably demonstrated. Room temperature curve analysis indicates a single broad (17-20 nm) gaussian, slightly asymmetric towards shorter wavelengths, in LHCII, CP26 and CP24 of PSII antenna [71] with maxima between 647–649 nm. In CP29 the chl*b* gaussian is substantially blue shifted to near 644 nm [71]. For LHCII the band asymmetry is significantly increased at 77 K [132] which may indicate the presence of vibrationally broadened spectral forms of chl*b*. Weak chl*b* emission has been recently demonstrated in isolated LHCII and thylakoids [133, 158] and a Stokes shift of about 5 nm has been suggested [133]. This value may need to be modified if chl*b* spectral forms are present.

5 Excitonic Energy Transfer

5.1 Singlet Energy Transfer Between Chlorophylls

Singlet excitonic energy transfer between chls is most commonly discussed in terms of the two limiting cases of very strong and very weak electronic coupling (J) between donor (D) and acceptor (A) transition dipoles [159–161]. J(cm^{-1}) may be calculated by the expression given by Pearlstein [162]

$$J = 5.04\, k \frac{\mu_D \cdot \mu_A}{R^3} \tag{2}$$

where μ_D and μ_A are the donor and acceptor transition moments (Debyes), k is a dipole orientation term and R (nm) is the D-A center to center distance.

In the weak coupling limit [160, 163] the transfer rate constant is given by

$$k_{DA} = \frac{4\pi^2\, |J|^2}{h\, \Delta\varepsilon} \tag{3}$$

where h is the Planck constant and $\Delta\varepsilon$ is the energy spread of the vibronic envelope of the electronic transition involved in energy transfer. This is the transfer mechanism named after Förster [160]. The more common form of the Förster expression, written in terms of experimentally accessible parameters, is

$$k_{\mathrm{DA}} = \frac{1}{\tau_0} \cdot \frac{R_0^6}{R^6} \tag{4}$$

where R_0 (nm) is defined by

$$R_0^6 = \frac{8.784 \times 10^{17}}{\mathrm{n}^4} k^2 \int \mathrm{d}\tilde{\nu}\, \tilde{\nu}^{-4}\, \mathrm{F_D}(\tilde{\nu})\, \varepsilon_{\mathrm{A}}(\tilde{\nu}) \tag{5}$$

The spectra $F_{\mathrm{D}}(\tilde{\nu})$ and $\varepsilon_{\mathrm{A}}(\tilde{\nu})$ are represented on the wavenumber scale and the fluorescence spectrum ($\mathrm{F}(\tilde{\nu})$) of the donor is normalized on this scale; n is the refractive index, $\varepsilon_{\mathrm{A}}(\tilde{\nu})$ is the molar decadic extinction coefficient of the acceptor and τ_0 is the radiative lifetime (s) and R(nm) is the D-A center to center distance. For very strong coupling the rate is given by

$$k_{\mathrm{DA}} = \frac{4|J|}{h} \tag{6}$$

More recently Knox and coworkers [163–165] have developed a unifying approach which explicitly takes into account exciton dephasing and which is applicable over all values of J. In this context the older ideas of very weak and very strong intermolecular electronic coupling are referred to the strength of vibrational interactions which broaden electronic transitions. These authors confirmed the applicability of the Förster mechanism to EET between chlorophylls (assuming the dominance of dipole-dipole interactions and after phase coherence is lost), even in the case of quite strong electronic coupling. This point is often overlooked in the photosynthesis literature. Thus it is possible to calculate pairwise EET rates if J is known (Eq. 3).

Recently a three-dimensional crystallographic structure with 6 Å resolution for the major antenna complex LHCII has been presented [51] and in which the 14–15 chls present are resolved (see Sect. 2). The plane of all chlorin rings appears to be almost perpendicularly oriented with respect to the major membrane plane though the geometrical molecular axes, and hence the transition dipole vectors, could not be determined. Nearest neighbour center to center chl-chl distances were estimated to be in the 0.9–1.4 nm range with most pigments having two neighbors. If one assumes an average, and rather favorable, orientation factor ($\mathrm{k} = 1$) and in vitro values for the chl transition moments [147, 166, 167], J can be determined from Eq. 2. The values for LHCII come out in the range $\mathrm{J} \approx 30$–$120\ \mathrm{cm}^{-1}$. A less favorable orientation term may be assumed for chl*b* as linear dichroism studies indicate that the $Q_y(0, 0)$ transition is oriented out of the membrane plane by more than 35° while the main long wavelength transitions lie close to the membrane plane (Sect. 4).

The chl-chl coupling estimated seems to be somewhat at variance with the suggestion of very strong coupling in LHCII, leading to delocalized, coherent excitonic interactions [51, 168].

Most attempts to calculate EET rates make use of the Förster mechanism together with in vitro chl data. R_0 for chl*a* (Eq. 5) is often taken as 90 Å [51, 169]. However as the published values of R_0 have a large spread [170] very large uncertainties arise in the rate calculations. We have therefore used the more general Förster expression (Eq. 4) to estimate pairwise transfer rates in LHCII. The values of J used are those estimated above and the energy spread ($\Delta\varepsilon$) of the chlorophyll Q_y transition was estimated to be about $2500\,\text{cm}^{-1}$. The pairwise EET rates come out between $4 \times 10^{11}\,\text{s}^{-1}$ and $6 \times 10^{12}\,\text{s}^{-1}$. As most chls are expected to be electronically coupled to more than one other chl the single step transfer rate ($\text{SSTT}^{-1} = n/\tau$ where τ is the pairwise transfer time and n is the number of nearest neighbors) will be higher than the pairwise values. From the crystallographic description of LHCII (Sect. 2) the single step transfer rate may reasonably be expected to be 2–3 times that of the pairwise rates. These values are on an average considerably lower than those suggested by Kühlbrandt and Wang for LHCII [51] for PSII antenna. Jia et al. [171], on the basis of estimates of the Förster overlap integrals for spectral forms in PSI using some in vitro chl parameters, calculated pairwise EET rates of the order of $10^{12}\,\text{s}^{-1}$ for $R = 11.5$ Å, in reasonable agreement with those determined above for LHCII using Eq. 4.

It should be pointed out that the Förster calculations are based on the point dipole assumption which may be inaccurate when the separation distance is similar to the molecular size, as is the case for LHCII. In this situation the transition monopole approximation should also be considered. For chl*a* Chang [172] has estimated that this leads to a Förster correction factor of 0.6–2.0 depending on orientation.

Pairwise EET rates cannot be directly measured in antenna systems. The closest approach to direct determination is offered on the one hand by time resolved picosecond and sub-picosecond absorption and fluorescence measurements and on the other hand by hole burning spectroscopies. Time resolved techniques do not detect transfer between isoenergetic sites. A somewhat more indirect approach to determining pairwise rates is that of analysing excited state lifetime data in terms of a particular antenna and an EET model.

In most cases time-resolved techniques have not to date provided very useful in determining pairwise EET rates between chl*a* molecules due to the pronounced wavelength overlap of the spectral forms. Also most studies have not been performed with the time resolution necessary to determine primary EET processes. There are several reports of sequential "downhill" energy transfer at low temperatures (4 K–77 K) between Chl*a* spectral forms [173–175] but with rather high time constants (100–200 ps). These presumably reflect excited state equilibration among pigment pools. Knox and Lin [176] have detected a 9–14 ps fluorescence decay component at low temperatures in PSII antenna, possibly associated with LHCII. Holzwarth et al. [177] and Lin et al. [178], using fluorescence and pump-probe techniques, describe a 10–14 ps component in *Synechococcus* PSI core antenna which is thought to be associated with equilibration between chl*a* pools. Lin et al. [178] have also demonstrated the presence of a faster component ($\tau < 5$ ps) which is probably due to energetically downhill spectral equilibration. In the narrow spectral range between 665 nm and 675 nm this equilibration seems approximately complete in less than 2 ps. This suggests that SSTT may be occurring on a picosecond or subpicosecond time scale for downhill energy transfer between energetically close chl*a* spectral forms. Such a conclusion may, however, be somewhat complicated by absorption associated with overlapping spectral forms. The low initial anisotropy values found by these authors at most wavelengths may be explained by absorption by overlapping spectral forms with different transition dipole orientations and may therefore not necessarily indicate extremely rapid (subpicosecond) EET. Decay associated fluorescence and absorption changes in the D1/D2/cytb559 complex, containing the RC of PSII suggest the presence of energy transfer processes with time constants ranging upward from 20 ps [179, 180]. Recently Durrant et al. [181] measuring transient absorption changes with femtosecond instrumental resolution provide evidence for ultra-fast excited state transfer (100 fs) between the accessory chlorin pool (chl plus pheophytin) absorbing near 670 nm and the 680 nm pool, thought to be dominated by the primary electron donor P680. Taking into account the stoichiometry of these pigment pools, pairwise transfer rates in the range $2\text{–}5 \times 10^{12}\ s^{-1}$ are calculated. The suggestion was made [181] that the slower transfer rates measured by others may represent only a small number of chromophores in the preparation. It should be mentioned however that the possibility cannot completely be excluded at the moment that the femtosecond absorption changes observed may in fact represent vibrational relaxation of excited states and not EET. The idea of subpicosecond transfer processes is supported by the recent observation that the 3 ps fluorescence component in the isolated D1/D2/cytb559 complex is from an approximately thermally equilibrated antenna state [182].

The relatively large spectral separation of chl*b* with respect to most of the chl*a* spectral forms has encouraged a number of attempts to determine SSTT by time-resolved techniques at room temperature. Thus Gillbro et al. [183] and Kwa et al. [184] have reported a 6 ± 4 ps chl*b* → chl*a* transfer time in isolated LHCII using pump-probe absorption techniques. The suggestion by Kwa et al. [184] of a faster (subpicosecond) component is difficult to evaluate in the light of a significant chl*a* absorption (30–35%) at the 650 nm pump wavelength used [133]. Working with a *Chlamydomonas reinhardtii* mutant containing only the chl*a*/*b* proteins of the external antenna of both photosystems (mostly LHCII), Eads et al. [185] suggest a chl*b* → chl*a* SSTT of about 0.5 ps. This yields an average chl*b* → chl*a* pairwise EET rate of around $10^{12}\ s^{-1}$ when the relative chl*a*-chl*b* stoichiometries are taken into account. Eads et al. [185] suggest that the longer (≈ 6 ps) times measured by others may be due to some uncoupling of chl*b* molecules during purification of LHCII. It should however be mentioned that chl*b* → chl*a* EET rates may in fact be considerably slower than "downhill" rates between most chl*a* spectral forms due to the less favourable Förster overlap integral. According to the calculations performed by Shipman and Housman [186], using chl solution parameters, this factor may lead to chl*b* → chl*a* transfer rates which are 2–3 times slower than between chl*a* spectral forms. In addition the dipole orientation term for transfer from chl*b* to chl*a* spectral forms do not seem to be very favorable (Sect. 4). For LHCII the possibility also exists that EET between isoenergetic chl*b* molecules may precede transfer to chl*a*. With these considerations in mind the measured time of around 6 ps may well be quite reasonable. Further investigation of this point with subpicosecond resolution seems desirable.

Hole burning spectroscopies have been quite extensively applied to chl-antenna systems in recent years. Narrow zero phonon lines have been observed in both absorption and fluorescence spectra [135–138] at low temperature (1.5–4 K), where it is thought that the line width is almost entirely due to excited state lifetime broadening [139, 140, 187]. From the extremely narrow holes burned in the absorption spectrum of PSI particles (FWHM $\approx 0.05\ \mathrm{cm}^{-1}$) extremely long SSTT of 200–400 ps have been estimated [139, 140]. A narrow hole ($0.85\ \mathrm{cm}^{-1}$) burned in the absorption spectrum of the PSII RC complex [136] suggests a 12 ps SSTT. Similar values (5–20 ps) may be calculated for the narrow holes in the fluorescence spectra of leaves [137, 138]. Hàla et al. [141] have recently reported the burning of a somewhat less narrow hole ($2\text{–}3\ \mathrm{cm}^{-1}$) in the emission spectra of PSII pigment-protein complexes, which yield SSTT values of around 4–5 ps. Pairwise transfer times are expected to be approximately 2–3 times slower. Thus most hole burning studies consistently indicate quite long transfer times. The suggestion has been made [140] that this may be caused by an increase in diagonal energy disorder together with decreased phonon-electron coupling at the low temperatures used in hole burning studies. This possibility, however, is not supported by the apparently rather weak temperature dependence between 35 K and 300 K demonstrated for the fluorescence lifetime component associated with trapping by PSI reaction centers [188].

Average pairwise EET rates for the inner antenna of PSI, containing only chl*a*, have been determined by Owens et al. [189, 190] by measuring the reaction center trapping time using time-resolved fluorescence techniques. The data were well described by the lattice model of Pearlstein [191] in which energy hops incoherently between isoenergetic lattice sites. Though the exact pairwise EET rate depends on the type of lattice model assumed, values come out quite close to $10^{12}\ \mathrm{s}^{-1}$.

In another model-based study Gillbro et al. [192] have used exciton-exciton annhilation to determine average EET hopping rates. In this technique the rate of exciton-exciton annhilation depends critically on the domain size and the pairwise EET rate [193, 194]. Average pairwise rates of $2 \times 10^{11}\ \mathrm{s}^{-1}\text{–}10^{12}\ \mathrm{s}^{-1}$ were calculated for LHCII having EET domain sizes in the range of 300–1000 sites.

It is clear from the experiments cited above that a bewildering range of EET rates have been reported for chl-chl transfer in photosynthetic antenna. Most of the long SSTT (> 10–20 ps) detected in time-resolved studies can probably be explained in terms of equilibration between pools of pigments or possibly between pigment protein complexes. Such an explanation does not however apply to most hole burning studies which for the time being are difficult to understand. Good agreement between the calculated pairwise transfer rates for LHCII using Förster theory ($4 \times 10^{11}\ s^{-1}$–$6 \times 10^{12}\ s^{-1}$) and the model dependent, average values for LHCII [192] and the core antenna of PSI [189, 190] are obtained. The time-resolved rate of about $10^{12}\ s^{-1}$ for pairwise chl*b* to chl*a* EET is also in good agreement with the calculated values and the larger SSTT of about 6 ps [183, 184] for this process may also not be in disagreement. For the D1/D2/cytb559 the time-resolved sub-picosecond EET rates of 2–$5 \times 10^{12}\ s^{-1}$ [181], while being somewhat greater than those determined for LHCII [192] and PSI core antenna [190], fall in the upper range of pairwise rates expected for the most strongly coupled chl molecules in LHCII. Thus the general impression which emerges from this survey is that there is quite good agreement between the Förster theory calculations and most experimental data pertinent to pairwise EET rates.

5.2 Singlet Energy Transfer from Carotenoids to Chlorophylls

Carotenoids are present in chl-protein complexes of higher plants (Sect. 2). The exact stoichiometry with respect to chl and the carotenoid-type varies in the different complexes [195] with overall levels being about 20% that of chl. Much higher carotenoid levels exist in some chl-containing algal photosystems. Triplet energy transfer from chl to carotenoid has long been known and has the important physiological role of protecting the photosynthetic apparatus from triplet chl [for review see 196]. Carotenoids are also known to have a light harvesting role and transfer singlet excitation to chl with efficiencies in the range of 60–100%, depending on the carotenoid and the organism [196–199].

Carotenoids are characterised by two low lying singlet excited states. The $S_2(^1B_u)$ state has a high absorption dipole strength and is very weakly fluorescent ($\Phi_F \approx 10^{-3}$–10^{-5}) [200, 201–204]. The $S_1(^2A_g)$ state is dipole forbidden and has been indirectly detected by transient S_1–S_n spectroscopy [206] and two photon fluorescence excitation spectroscopy [207]. Recently clear demonstrations of S_1 emission have been published for a number of carotenoids in vitro [204, 205, 208]. The Φ_F is low and the excited state lifetime is in the 15–60 ps range, somewhat longer than for only S_2 emitters. The relationship between chemical structure and S_1 emission has been discussed by Mimuro et al. [204] who suggested this to be associated with 8 conjugated double bonds and a nearby keto oxygen. Evidence has been recently presented for extremely weak one photon S_1 absorption by two bacterial carotenoids. The extinction coefficient is about 0.04% that of the dipole allowed S_0–S_2 transition [209]. It seems reasonable to expect that in the near future S_1 absorption will be observed for other carotenoids.

A detailed understanding of singlet car-chl EET is not yet available. Albrecht and colleagues have determined some transfer times in several algal systems using pump-probe absorption techniques with femtosecond resolution [199, 205]. The results demonstrate very fast SSTT values in the range 240 fs–2 ps, indicating reasonably strong electronic coupling between the two kinds of chromophore. From an estimate of emission-absorption overlap and measured transfer times it has been suggested that car-chl coupling energies may be of the order of 60–120 cm^{-1} [199].

The most widely held view concerning car-chl interactions is that exchange coupling [161] dominates [169, 196, 210, 211]. This idea was developed to explain efficient energy transfer from car donors which have extremely low fluorescence yields [210] and assumes that the shortest car-chl distance does not exceed the van der Waals contact distance ($\approx$ 4 Å). The transfer rate is given by:

$$k_{DA} = \frac{4\pi^2}{h}(J_{DA}^{ex})^2 \int d\nu\, F_D(\nu)\, \varepsilon_A(\nu) \tag{7}$$

where F_D is the normalized fluorescence emission spectrum of the donor, ε_A is the normalized absorption spectrum of the acceptor while J_{DA}^{ex} has the dimension of energy and cannot be associated with optical parameters. It should be emphasised that $\varepsilon_A(\nu)$ is not the concentration absorption spectrum and also that the radiative lifetime term, present in the Förster expression (Eq. 4), is here absent. Thus the exchange mechanism, which contains substantially the same spectral overlap term as the Förster mechanism (Eq. 5) does not depend on donor/acceptor dipole strengths and was therefore considered particularly attractive to explain EET from carotenoids to chls. In this reasoning the low carotenoid fluorescence yield seems to have been erroneously equated with low dipole strength and this seems to have given rise to some misconceptions. In fact, available evidence indicates that the radiative lifetimes (τ_0) for carotenoids are quite short e.g. S_2–S_0 β-carotene, $\tau_0 \approx 1$ ns [200]; S_2–S_0 fucoxanthin, $\tau_0 \approx 3$ ns [205]; S_1–S_0 fucoxanthin, $\tau_0 \approx 50$ ns [205], indicative of quite high dipole strengths, which are comparable with those of chl ($\tau_0 \approx 15$ ns). Thus while the exchange mechanism may very well be operative in car-chl EET, Coulombic interactions should definitely not be excluded, as pointed out recently by Shreve et al. [200, 205] and Owens et al. [207].

Since recognition that antenna carotenoids have a low lying S_1 electronic state it has been generally assumed that EET to chl proceeds from this state. This is based largely on two considerations. 1. Spectral overlap between the carotenoid S_1 state and the Q_y chl absorption transition is more favourable than for the S_2 state. This assumption is however not supported by the recent study of Mimuro et al. [204] in which spectral overlap can be shown to be rather similar for a number of mainly S_1 and mainly S_2 emitters. This point may therefore need further clarification. 2. Extremely fast S_2–S_1 internal conversion. Recent studies based on fluorescence yield measurements and also on femtosecond absorption measurements [200, 205, 207] indicate in vitro S_2–S_1 relaxation times of

150–300 fs. The measured in vitro S_1 lifetimes are much longer and have been determined in the range 10–60 ps [200, 204, 205].

Several comments on points 1 and 2 may be made. Concerning the spectral overlap between carotenoid emission and chl Q_y absorption one notes, on the basis of recently published S_1 emission spectra [204], that spectral overlap with chl absorption is not extremely favorable as emission maxima are red-shifted by 70–80 nm with respect to the Q_y chl absorption band. As the carotenoid electronic state energies are shifted to lower values with increasing number of conjugated double bonds [212, 213] S_1 overlap with chl may be even less favorable for such important higher plant carotenoids as lutein, β-carotene and zeaxanthin which have longer conjugated systems than the S_1 emitters so far examined. In these cases it may well be necessary to consider the S_2 states in energy transfer discussions, as suggested by Shreve et al. [205]. In this context it should be noted that the measured in vitro S_2–S_1 internal conversion times are not very much shorter than the in vivo car-chl SSTT values determined for some algal carotenoids, thus suggesting the possibility that EET from the carotenoid S_2 state may in some cases be competitive with S_2–S_1 internal conversion [207].

5.3 Singlet Energy Transfer from Chlorophylls to Carotenoids

As mentioned above, triplet excitation is known to be transferred from chl to carotenoids. This proceeds via exchange coupling and has an important physiological protective role. Singlet EET from chl to carotenoid has however been little discussed in the photosynthetic literature due to the apparently poor spectral overlap with chl emission ($Q_y(0,0)$ emission maximum near 683 nm). Extremely rapid EET between antenna chls (Sect. 5.1) would have rendered singlet transfer to carotenoids extremely uncompetitive. In addition this process seemed physiologically uninteresting as EET to RCs involves chl singlet states. In recent years, several lines of investigation have somewhat changed this scenario.

Firstly, as pointed out in Sect. 2, the carotenoid zeaxanthin has been implicated in a reversible quenching process (qE) in PSII which protects RCs from high-light induced photoinhibition. Whilst other possibilities exist to mechanistically explain this quenching process [214] one attractivly simple hypothesis is that singlet energy is transferred from antenna chl to zeaxanthin where it is thermally dissipated. Secondly the apparently insurmountable problem associated with carotenoid-chl spectral overlap has been greatly redimensioned in recent years by the demonstration that S_0–S_1 transitions are at considerably lower energies [207, 209] than was previously thought [203, 215–218]. Thus Owens et al. [207] place the S_0–S_1 transition of fucoxanthin (8 conjugated C–C double bonds plus a keto oxygen) near 620 nm on the basis of two photon excitation spectra. Single photon absorption measurements

[209] place this transition near 623 nm and 659 nm for two bacterial carotenoids (neurosporene, 9 conjugated C–C double bonds; spheroidene, 10 conjugated C–C double bonds).

If zeaxanthin is able to quench chl singlet excitation, its S_0–S_1 transition should lie above 670 nm in order achieve good spectral overlap with chl emission. Experimental data for the S_0–S_1 transition of zeaxanthin or other carotenoids with 11 C–C conjugated double bonds is lacking. However it is known from studies with carotenoids having smaller conjugated systems that electronic transitions shift to lower energies with increasing numbers of conjugated double bonds [212, 213], thus suggesting the possibility that the S_0–S_1 transition of zeaxanthin may be located between 660–690 nm [206, 213].

Owing to the forbidden, or at least very weak, nature of S_0–S_1 absorption, EET to carotenoids from chl is expected to be via the exchange coupling mechanism [161], which does not require donor/acceptor dipole strength. In this case the chl-car distance would need to be very short, i.e. less than the van der Waals contact distance (≈ 4 Å).

6 Models for Photosystem Antenna

6.1 Funnel Model

Since the discovery of the chl spectral forms in the early 1970s it has often been suggested that they are organized spatially with respect to RCs. Thus shorter wavelength forms were envisaged to be concentrated in the peripheral antenna with the longer wavelength forms close to RCs. In this way spectral heterogeneity was rationalised in terms of an energy sink in which excitation energy was directed towards RCs [186, 219, 220]. It was demonstrated that such an organisation increases the rate of EET to RCs many times. A number of studies in recent years however suggest that for higher plant photosystems this concept should be re-examined.

For PSI core antenna, an extensive series of analysis have been performed by Fleming and coworkers, using core particles containing different chl/P700, ratios, which indicate that the "funnel" organization is not applicable to this antenna system. It is shown: (a) that the steady state absorption properties are similar for particles with different sizes [189]. (b) the RC-dependent trapping time (15–40 ps) is linearly related to antenna size [221]. (c) The spectroscopic properties of the RC-trapping fluorescence decay component (15–40 ps) are rather similar in core particles with different antenna sizes and does not demonstrate a rapid concentration of excited states in long wavelength forms as is expected for the funnel organization [189]. (d) the trapping time shows little temperature dependence [188] in marked contrast to model calculation predictions of the "funnel" structure [171].

Precise information is lacking on whether the "funnel" organization may play a significant role in EET from the peripheral to the core antenna of PSI. Owens et al. [190] suggest that this process might be extremely rapid, even faster than trapping within the core (15–40 ps), as no rising component could be detected when excitation was partially into the peripheral antenna with an instrumental resolution time of about 10 ps. If this interpretation is correct, it could imply an ordered (funnel) arrangement of spectral forms which allows rapid EET from the peripheral to the core antenna.

For PSII antenna, some disagreement exists in the literature on the distribution of the main long wavelength spectral form. The presence of significant levels of the 684 nm form in the main outer antenna complex was demonstrated in room temperature absorption and fluorescence spectra [71, 132]. This conclusion contrasts with that of van Dorssen et al. [131] who suggested that almost all the 684 nm forms were localized in the core antenna. Hemelrjck et al. [156] subsequently failed to find evidence for long wavelength electronic transitions in isolated LHCII at cryogenic temperatures. These latter spectroscopic studies were, however, performed at low temperatures, which have been shown to drastically reduce the amount of the 684 nm form in LHCII [132].

It has been suggested that excitation energy is not concentrated in the core antenna of PSII as the overall steady state emission spectrum of this PS is very similar to that of isolated LHCII and differs from that of the core antenna [158]. Very recently the distribution of the spectral forms in all six antenna chl-protein complexes of PSII (Sect. 2) was described [71]. Outer and inner antenna complexes were shown to contain the same spectral forms in rather similar proportions. Using an equilibrium thermodynamic approach the free energy (ΔG_0) for exciton transfer from outer to core antenna was calculated to be around $-0.17\ \text{kcal mol}^{-1}$, considerable less than RT. The free energy change for EET between the two main core antenna complexes CP47 and CP43 was close to zero. It was concluded that PSII is organized as a very shallow funnel, which is entirely due to the presence of chl*b* in the outer antenna complexes. The chl*a* spectral forms in the outer and core antenna are on an average isoenergetic. A recent analysis of the steady state emission of the six antenna complexes plus the RC complex lends experimental support to this conclusion [72].

It has been noted in recent years that all antenna systems seem to contain minor spectral forms which absorb at somewhat lower energies than RCs. This has led to the suggestion that they may be closely associated with RCs and could thus increase trapping rates by focusing energy on the RCs in photosynthetic bacteria [222–225] and higher plants [226–228]. Some support for the presence of long wavelength spectral forms near RCs of PSI comes from model studies of Jia et al. [171]. These authors found that placing a small number of low energy pigments near RCs, in an antenna bed with a homogeneous distribution of the main spectral forms, was necessary to simulate the temperature and wavelength dependence of the trapping rate in PSI core particles [188]. This conclusion is, however, subject to the model assumptions made and several of these may be questioned. Firstly the in vitro chl band characteristics were assumed while there is good reason to believe that these differ somewhat from those of the spectral forms (see Sect. 4). Secondly the fluorescence Stokes shifts were assumed to be insensitive to temperature. This is unrealistic as the absorption band width was calculated to change markedly with temperature which would lead to considerable decreases in the Stokes shift [157]. It is not clear to what extent these simplifying assumptions might affect the calculated results. It should also be noted that the long wavelength forms considered by Jia et al. [171] are to be distinguished from those analysed by Mukerji and Sauer [228] which occur in the peripheral antenna complex LHCI.

For PSII it has been demonstrated by gaussian band analysis of room temperature absorption spectra that long wavelength bands are not concentrated in the chl-protein complexes of the core antenna [71].

Model kinetic studies have cast serious doubt on the utility for trapping of low energy chlorophylls near RCs [229]. It is shown that only when the low energy antenna component is isoenergetic (or nearly so) with the RC can increased trapping rates ensue. In the case of large energy differences (> kT) decreased trapping rates are expected. For PSI and PSII the main long wavelength antenna forms are at 705 nm and 684 nm respectively which means that the energy differences with respect to RCs is around 0.5 kT. Thus if they were concentrated near RCs one might at best expect a modest increase in the trapping rate. Some evidencc cxists for very low levels of even longer wavelength bands in the low energy absorption tail of PSI [230] and PSII [71, 132] for which the energy difference with respect to RCs is in the range of 1–2 kT. The biologically important parameter is, however, trapping efficiency and small changes in trapping rate will not lead to significant modifications in trapping efficiency [229, 231]. Thus it seems likely that the low energy forms in higher plant photosystems may not have an important role in energy trapping.

6.2 Trapping vs Diffusion Limited Models

An important question concerning energy trapping is whether its kinetics are limited substantially by (a) exciton diffusion from the antenna to RCs or (b) electron transfer reactions which occur within the RC itself. The former is known as the diffusion limited model while the latter is trap limited. For many years PSII was considered to be diffusion limited, due mainly to the extensive kinetic modelling studies of Butler and coworkers [232, 233] in which this hypothesis was assumed. More recently this point of view has been strongly contested by Holzwarth and coworkers [230, 234, 235] who have convincingly analyzed the main open RC PSII fluorescence decay components (200–300 ps, 500–600 ps for PSII with outer plus inner antenna) in terms of exciton dynamics within a system of first order rate processes. A similar analysis has also been presented to explain the two PSII photovoltage rise components (300 ps, 500 ps) by Liebl et al. [236]. In this model the faster component represents trapping associated with primary charge separation in a trap limited PSII in which energy flows in and out of P680 many times (10–20 times) before trapping occurs. The slower component is largely determined by charge stabilization at the level of the primary quinone acceptor and therefore proposes that primary charge separation is readily reversible. In this model the antenna-RC is considered as a single species. Thermodynamic equilibration between antenna chlorophylls and P680 (P-A equilibration) is assumed to be extremely fast (< 10–20 ps) [142, 235]. However in view of the calculated Förster pairwise transfer rates for LHCII (Sect. 5) this proposed super-fast equilibration may be legitimately questioned. In a random walk process of exciton migration in a regular lattice structure, the average number of hopping steps prior to arriving at the RC site is thought to be slightly greater or equal to the number of pigment sites within the antenna matrix [237] i.e. about 250 for PSII (Sect. 2). If it is assumed that excitation transfer between pigment-protein complexes has approximately similar dynamics to that occurring within each complex, P-A equilibration might therefore be expected to take not less than 50–100 ps. Experimentally P-A equilibration would be extremely difficult to observe by measurements of picosecond decay associated spectra as the chl*a* absorption and fluorescence emission spectra of all pigment-protein complexes comprising the PSII antenna are very similar [71, 72] and P680 is expected to introduce only a slight spectral pertubation [158]. Thus equilibration within each monomer chl-protein complex, which might reasonably be expected to occur in less than 10–20 ps, would be almost indistinguishable spectroscopically from P-A equilibration. As pointed out in Sect. 2, each PSII contain 17–22 separate chl-binding polypeptides. This clearly means that the apparent, spectroscopically determined equilibration time may be much shorter than P-A equilibration time and thus less than the time required for excitation to arrive at P680.

An important prediction of this trap limited model is that primary charge separation is reversible and that the radical pair recombination time should be less than 1 ns. Concerning the direct measurement of this parameter, there is to date little agreement in the literature on this point. Most transient absorption studies suggest slow recombination kinetics which go from 4 ns [238] and 11 ns [239, 240] for PSII core particles to tens of nanoseconds [241, 242]. The possibility has been suggested that these long lifetime recombination reactions may be due to artificially modified RCs [238]. These authors present data which are interpreted to indicate quite high yields of an experimentally unresolved subnanosecond recombination component. Schatz et al. [234] have described a 1–2 ns transient absorption and fluorescence component which may be associated with subnanosecond charge recombination. This interesting point therefore remains to be definitively established.

For PSI core it has been argued that excitation visits P700 2–3 times on an average prior to being trapped [221]. This conclusion is based on the analysis of time resolved fluorescence trapping times as a linear function of antenna size in terms of the Pearlstein model array treatment (see Sect. 5). A similar conclusion has recently been published for Synechococcus PSI core particles based on a global analysis of fluorescence decays [243]. In this study spectral equilibration was shown to be complete in about 12 ps with a RC trapping time of 35 ps. It was therefore suggested that excitation visits P700 about 3 times prior to trapping. If, as suggested above for PSII, spectral equilibration in PSI is faster than the first RC passage time, this value may then be viewed as an upper limit. Exciton dynamics in PSI would therefore seem not to be determined exclusively be either the trap or diffusion limited extremes, but by a situation in which both processes are important.

An important question which is relevant to the discussion on whether trapping is diffusion or trap limited is the overall trapping time (τ_t). It is generally accepted that this process is much slower in PSII than in PSI [169, 236, 244–246], with (PSII) 300 ps and (PSI) 50–100 ps. Schatz et al. [235] have shown that assuming a Boltzmann distribution of excited states between RCs and an N-fold degenerate antenna (N is the number of antenna sites) a very reasonable numerical relationship is determined between τ_t and the reaction time for pheophytin reduction ($\tau_r \approx 2.8$ ps). Applying this reasoning to PSI and assuming similar values for N and the primary photochemical reaction time to those for PSII [221, 247] the τ_t(PSI)/τ_t(PSII) ratio is about 7. This value is not very far from the ratio of the measured values which are in the range 3–6. These differences in overall trapping time calculated for the two photosystems are due to the fact that P680 is less than 0.5 kT below its antenna while P700 is about 2 kT lower. Thus it is not necessary to invoke any large structural or functional difference at the antenna level to explain the large difference in trapping rate of the two photosystems. This small number of exciton visits to PSI RCs suggested by Owens et al. [221] and Turconi et al. [243] is probably quite sufficient to establish equilibrium with the antenna. Thus this analysis does not depend on the extreme trap limited model.

7 Physiological Importance of Slow Trapping in Photosystem II

As discussed above (Sect. 6) the RC-trapping time for PSII is extremely long ($\tau_t > 300$ ps) compared with PSI. This long trapping time is associated with (a) P680 being almost isoenergetic with the bulk of the chl*a* antenna and (b) weak energy "funnelling". It has very recently been proposed that this slow RC-trapping may have a precise physiological function in the down-regulation of excited state levels within PSII at increasing light intensities [72]. This process, caused by the creation of trapping centers within the PSII antenna at high absorbance fluxes [72, 248–250], is important in protecting RCs from photoinhibition. Such trapping efficiency (Φ_T) in an EET matrix is expected to be linear with the RC-trapping time (τ_t) according to $\Phi_T = k_T \cdot \tau_t$, where k_T is the rate of trapping by antenna trapping centers. Thus it is clear that a long τ_t will favour excited state down-regulation. These antenna trapping centers seem to be localized, at least in part, in the outer antenna of PSII [72, 248–250]. As pointed out by Jennings et al. [72] such protective mechanisms would be scarcely feasible if the antenna was organized as a deep energy "funnel" as excitons would be rapidly transferred to the core antenna, with a subsequent low "escape" probability to the outer antenna.

8 References

1. Melis A, Anderson JM (1983) Biochim. Biophys. Acta 724: 473
2. McCauley SW, Melis A (1986) Biochim. Biophys. Acta 849: 175
3. Mauzerall D, Greenbaum NL (1989) Biochim. Biophys. Acta 974: 119
4. Cline K (1986) Biol. Chem. 261: 10804
5. Simpson DJ (1978) Carlsberg Res. Commun. 43: 365
6. Miller KR, Kushman RA (1979) Biochim. Biophys. Acta 546: 481
7. Berthold DA, Babcock GT, Yocum CF (1981) FEBS Lett. 134: 231
8. Bassi R, Hoyer-Hansen G, Barbato R, Giacometti GM, Simpson DJ (1987) J. Biol. Chem. 262: 13333
9. Peter GF, Thornber JP (1991) J. Biol. Chem. 266: 16745
10. Dainese P, Bassi R (1991) J. Biol. Chem. 266: 8136
11. Satoh K (1985) Photochem. Photobiol. 42: 845
12. Namba O, Satoh K (1987) Proc. Natl. Acad. Sci. USA 84: 109
13. Morris J, Herrman RG (1984) Nucleic Accids Res. 12: 2837
14. Vermaas WFJ, Williams JGK, Arntzen CJ (1987) Plant Mol. Biol. 8: 317
15. Holschuhl K, Bottomley W, Whitefeld PR (1984) Nucleic Acid Res. 12: 8819
16. Bricker TM (1990) Photosynth. Res. 24: 1
17. de Vitry C, Wollmann FA, Delepelaire P (1984) Biochim. Biophys. Acta 767: 415
18. Glick RE, Melis A (1989) Biochim. Biophys. Acta 934: 151
19. Barbato R, Race HL, Friso G, Barber J (1991) FEBS Lett. 286: 86
20. Marquardt J, Bassi R (1993) Plantal 191: 265
21. Thornber JP, Smith CA, Bayley JL (1966) Biochem. J. 100: 14
22. Butler PJG, Kühlbrandt W (1988) Proc. Natl. Acad. Sci. USA 85: 3797
23. Bassi R, Pineau B, Dainese P, Marquardt J (1993) Eur. J. Biochem. 212: 297
24. Di Paolo ML, Peruffo dal Belin A, Bassi R (1990) Planta 181: 275
25. Spangfort M, Andersson B (1989) Biochim. Biophys. Acta 977: 163
26. Sigrist M, Staehelin LA (1992) Biochim. Biophys. Acta 1098: 191
27. Larsson UK, Sundby C, Andersson B (1987) Biochim. Biophys. Acta 894: 59
28. Bassi R, Giacometti GM, Simpson D (1988) Biochim. Biophys. Acta 935: 152
29. Dunsmuir P (1985) Nucleic Acids Res. 13: 2503
30. McGrath JM, Terzaghi WB, Sridhar P, Cashmore AR, Pichersky E (1991) Plant. Mol. Biol. 19: 725
31. Jansson S, Pichersky E, Bassi R, Green BR, Ikeuki M, Melis A, Simpson DJ, Spangfort M, Staehelin LA, Thornber JP (1992) Plant Mol. Biol. Reporter 10: 242
32. Mattoo AK, Edelman M (1987) Proc. Natl. Acad. Sci. USA 84: 1497
33. Bennett J (1977) Nature 269: 344
34. Lamppa GK, Abad MS (1987) J. Cell Biol. 105: 2641
35. Jansson S, Seltman E, Gustafsson P (1990) Biochim. Biophys. Acta 1019: 110
36. Jansson S, Gustafsson P (1994) Plant. Mol. Biol. in press
37. Kohorn BD, Harel E, Chitnis PR, Thornber JP, Tobin EM (1986) J. Cell Biol. 102: 972
38. Green B, Pichersky E, Kloppstech K (1991) T.I.B.S. 16: 181
39. Wedel N, Klein R, Ljunberg U, Andersson B, Herman RG (1992) FEBS Lett. 314: 61
40. Kim S, Sandusky P, Bowlby NR, Aebersold R, Green BR, Vlahakis S, Yocum CF, Pichersky E (1992) FEBS Lett. 314: 67

41. Thornber JP, Peter GF, Morishige DT, Gomez S, Anandan S, Kerfeld C, Welty BA, Lee A, Takeuki TS, Preiss S (1993) Biochem. Soc. Trans. 21 in press
42. Allen JF, Bennet J, Steinbeck KE, Arntzen CJ (1981) Nature 291: 25
43. Allen JF (1992) Biochim. Biophys. Acta 1098: 275
44. Telfer A, Bottin H, Barber J, Mathis P (1984) Biochim. Biophys. Acta 764: 324
45. Horton P (1983) FEBS Lett. 152: 47
46. Jennings RC, Islam K, Zucchelli G (1986) Biochim. Biophys. Acta 850: 483
47. Forti G, Vianelli A (1988) FEBS Lett. 231: 95
48. Bassi R, Rigoni F, Barbato R, Giacometti GM (1988) Biochim. Biophys. Acta 936: 29
49. Bassi R, Dainese P (1992) In: Argyroudy-Akoyonoglou J (ed) Regulation of chloroplast development, Plenium Press New York London p 511
50. Wang DN, Kühlbrandt W (1991) J. Mol. Biol. 217: 691
51. Kühlbrandt W, Wang DN (1991) Nature 350: 130
52. Broglie R, Bellamere G, Barlett SG, Chua NH, Cashmore AR (1981) Proc. Natl. Acad. Sci. USA 78: 7304
53. Coruzzi G, Broglie R, Cashmore AR, Chua N-H (1983) J. Biol. Chem. 258: 13399
54. Buetow DE, Chen H, Erdos U (1988) Photosynth. Res. 18: 61
55. Spangfort M, Larsson UK, Ljunberg U, Ryberg M, Andersson B (1990) In: Baltscheffsky M (ed) Current Research in Photosynthesis, Kluwer Academic Publishers, Dordrecht, vol 2 p 253
56. Michel HP, Buvinger WE, Bennett J (1990) In: Baltscheffsky M (ed) Current research in photosynthesis, Kluwer Dordrecht, vol 2 p 747
57. Mullett JE (1983) J. Biol. Chem. 258: 9941
58. Larsson UK, Anderson B (1985) Biochim. Biophys. Acta 809: 396
59. Islam K (1987) Biochim. Biophys. Acta 893: 333
60. Maempaa P, Andersson B (1989) Z. Naturforschnung 44c: 403
61. Morishige D, Thornber JP (1991) FEBS Lett. 293: 183
62. Schwartz E, Stasys R, Aebersold R, McGrath JM, Green BR, Pichersky E (1991) Plant Mol. Biol. 17: 923
63. Bassi R, Dainese P (1991) Eur. J. Biochem. 204: 317
64. Tremoliere A, Dainese P, Bassi R (1993) Eur. J. Biochem. in press
65. Demmig-Adams B (1990) Biochim. Biophys. Acta 1020: 1
66. Machold O, Simpson DJ, Moller BL (1979) Carlsberg Res. Commun. 44: 235
67. Camm EL, Green BR (1980) Plant Physiol 66: 428
68. Henrysson T, Schroeder WP, Spangfort M, Akerlund HE (1989) Biochem. Biophys. Acta 977: 301
69. Dainese P, Hoyer-Hansen G, Bassi R (1990) Photochem. Photobiol. 51: 693
70. Barbato R, Rigoni F, Giardi MT, Giacometti GM (1989) FEBS Lett. 251: 147
71. Jennings RC, Bassi R, Garlaschi F, Dainese P, Zucchelli G (1993) Biochemistry 32: 3203
72. Jennings RC, Garlaschi FM, Bassi R, Zucchelli G, Vianelli A, Dainese P (1993) Biochim. Biophys. Acta 1183, 194
73. Irrgang K-D, Renger G, Vater J (1991) Eur. J. Biochem. 201: 515
74. Hayden DB, Baker NR, Percival MP, Beckwith PB (1987) Biochim. Biophys. Acta 851: 86
75. Bassi R, Simpson DJ (1986) Carlsberg Res. Commun. 51: 363
76. Dunahay TG, Shuster G, Staehlin LA (1987) FEBS Lett. 215:25
77. Allen KD, Staehelin LA (1992) Plant Physiol. 100: 1517
78. Pichersky E, Subramaniam R, White MJ, Reid J, Aebersold R, Green B (1991) Mol. Gen. Genet. 227: 277
79. Sorensen AB, Lauridsen BF, Gausing K (1992) Plant Physiol. 98: 1538
80. Arvidson PO, Bratt CE, Andreasson L-E, Akerlund H-E (1992) In: Murata N (ed) Research in photosynthesis. Kluwer, Dordrecht, vol 1 p 235
81. Morishige DT, Anandan S, Jaing JI, Thornber JT (1990) FEBS Lett. 264: 242
82. Schwartz E, Pichersky E (1990) Plant Mol. Biol. 15: 157
83. Bassi R, Soen SY, Frank G, Zuber H, Rochaix JD (1992) J. Biol. Chem 267: 25714
84. Krishnan M, Gnanam A (1979) FEBS Lett. 97: 322
85. Irrgang KD, Bochtel C, Vater J, Renger G (1990) In: Baltscheffsky M (ed) Current research in photosynthesis. Kluwer, Dordrecht, vol 1 p 355
86. Grimm B, Kruse E, Kloppstech K (1989) Plant Mol. Biol. 13: 583
87. Adamska I, Ohad I, Kloppstech K (1992) In: Argyroudy-Akoyonoglou JH (ed) Regulation of chloroplast biogenesis. Plenum New York, p 113
88. Hundal T, Virgin I, Styring S, Andersson B (1990) Biochim. Biophys. Acta 1017: 235

89. Bassi R, Ghiretti Magaldi A, Tognon G, Giacometti GM, Miller K (1989) Eur. J. Cell. Biol. 50: 84
90. Boekema EJ, Wynn RM, Malkin R (1990) Biochim. Biophys. Acta 1017: 49
91. Mörshel E, Schatz GH (1987) Planta 172: 145
92. Dekker JP, Bowlby NR, Yocum CF (1989) FEBS Lett 254: 150
93. Siebert M, De Witt M, Staehlin LA (1987) J. Cell. Biol. 105: 2257
94. de Vitry C, Diner BA, Popot JL (1991) J. Biol. Chem. 266: 16614
95. de Vitry C, Wollman FA, Delepelaire P (1983) C.R. Acad. Sci. Paris 279: 277
96. Holzenburg A, Bewley MC, Wilson FH, Nicholson WV, Ford RC (1993) Nature 363: 470
97. Simpson DJ (1979) Carlsber Res. Commun. 44: 305
98. Lyon MK, Miller KR (!985) J. Cell. Biol. 10: 1139
99. Santini C, Tidu V, Tognon G, Ghiretti-Magaldi A, Bassi R (1994) Eur. J. Biochem. (in press)
100. Harrison MA, Melis A (1992) Plant Cell Physiol. 33: 627
101. Morrissey PJ, Glick RE, Melis A (1989) Plant Cell Physiol. 30: 335
102. Camm EL, Green BR (1980) Biochim. Biophys. Acta 974: 180
103. Tidu V (1993) Thesis: Dipartimento di Biologia, Università di Padova Italy
104. Simpson DJ (1983) Eur. J. Cell Biol. 31: 305
105. Mullett JE, Burke JJ, Arntzen CJ (1980) Plant Physiol. 65: 814
106. Nechustai R, Peterson CC, Peter GF, Thornber JP (1987) Eur. J. Biochem. 164: 345
107. Bassi R, Machold O, Simpson DJ (!985) Carlsberg Res. Comm. 50: 145
108. Peter G, Takeuchi T, Thornber JP (1991) Methods: a Companion to Methods Enzymol. 3: 115
109. Bruce ED, Malkin R (1988) J. Biol. Chem. 263: 7302
110. Goldbeck JH (1987) Biochim. Biophys. Acta 895: 167
111. Lam E, Ortiz W, Malkin R (1984) FEBS Lett. 168: 10
112. Wollman FA, Bennoun P (1982) Biochim. Biophys. Acta, 680: 352
113. Ikeuki M, Hirano A, Inoue Y (1991) Plant Cell Physiol. 32: 103
114. Knoetzel J, Svendsen I, Simpson DJ (1992) Eur. J. Biochem. 206: 209
115. Hoffman NE, Pichersky E, Malik VS, Castresana C, Ko K, Darr SC, Cashmore AR (1987) Proc. Natl. Acad. Sci. USA 84: 8844
116. Jensen PE, Kristensen M, Hoff T, Lehnbeck J, Stumman BM, Henningsen KW (1992) Physiol. Plant. 84: 561
117. Palomares R, Herrmann RG, Oelmüller R (1991) J. Photochem. Photobiol. B: Biol. 11: 151
118. Zhang H, Hanley S, Goodman HM (1991) Plant Physiol. 96: 1387
119. Damm I, Steinmetz D, Grimme LH (1990) In: Baltscheffsky M (ed) Current research in photosynthesis. Kluwer Dordrecht, vol 2 p 607
120. Preiss S, Peter G, Anandan S, Thornber JP (1993) Photochem. Photobiol. 57: 152
121. Knoetzel J, Simpson DJ (1991) Planta 185: 111
122. Stayton MM, Brosio P, Dunsmuir P (1987) Plant Mol. Biol. 10: 127
123. Pichersky E, Tanksley SD, Piechulla B, Stayton MM, Dunsmuir P (1988) Plant Mol. Biol. 11: 69
124. Pichersky E, Brock TG, Nguyen D, Hoffman NE, Piechulla B, Thanksley S, Green BR (1989) Plant Mol. Biol. 12: 257
125. Steppuhn J, Hermans J, Nechushtai R, Ljumberg F, Thumler F, Lottspeich F, Hermann R (1988) FEBS Lett 237: 218
126. Hayes JM, Gillie JK, Tang D, Small GJ (1988) Biochim. Biophys. Acta 932: 287
127. Ormos P, Ansari A, Braunstein D, Cowen BR, Frauenfelder H, Hong MK, Iben IET, Sanke TB, Steinbach PJ, Young RD (1990) Biophys. J. 57: 191
128. French CS, Brown JS, Lawrence MC (1972) Plant Physiol. 49: 421
129. Brown JS, Anderson JM, Grimme L (1982) Photosynth. Res. 3: 279
130. Brown JS, Schoch S (1982) Photosynth. Res. 3:19
131. van Dorssen RJ, Plijter JJ, Dekker JP, den Ouden A, Amesz J, van Gorkom HJ (1987) Biochim. Biophys. Acta 890: 134
132. Zucchelli G, Jennings RC, Garlaschi FM (1990) J. Photochem. Photobiol., B: Biol. 6: 381
133. Zucchelli G, Jennings RC, Garlaschi FM (1992) Biochim. Biophys. Acta 1099: 163
134. Leupold D, Stiel H, Hoffmann P (1988) In: Scheer H, Schneider S (eds) Photosynthetic light-harvesting systems-Organisation and function: proceedings of an international workshop, 12–16 October 1988. Freising, Germany. Walter de Gruyter, Berlin, P 387
135. Vacha M, Adamec F, Ambroz M, Baumruk V, Dian J, Nedbal L, Hala J (1991) Photochem. Photobiol. 54: 127

136. Tang D, Jankowiak R, Seibert M, Yocum CF, Small GJ (1990) J. Phys. Chem. 94: 6519
137. Renge I, Mauring K, Vladkova R (1988) Biochim. Biophys. Acta 935: 333
138. Mauring K, Renge I, Avarmaa R (1987) FEBS Letters 223: 165
139. Gillie JK, Hayes JM, Small GJ, Golbeck JH (1987) J. Phys. Chem. 91: 5524
140. Gillie JK, Small GJ, Golbeck JH (1989) J. Phys. Chem. 93: 1620
141. Hála J, Vácha M, Dian J, Prásil O, Komenda J (1992) Photosynthetica 26: 429
142. McCauley SW, Bittersmann E, Holzwarth AR (1989) FEBS Letters 249: 285
143. Jennings RC, Zucchelli G, Garlaschi FM (1990) Biochim. Biophys. Acta 1016: 259
144. Ikegami I, Itoh S (1986) Biochim. Biophys. Acta 851: 75
145. van Ginkel G, Kleinen Hammans JW (1980) Photochem. Photobiol. 31: 385
146. Gudowska-Nowak F, Newton MD, Fajer J (1990) J. Phys. Chem. 94: 5795
147. Shipman LL, Cotton TM, Norris JR, Katz JJ (1976) J. Am. Chem. Soc. 98: 8222
148. Brunisholz RA, Zuber H (1992) J. Photochem. Photobiol. 15: 113
149. Fowler GJS, Visschers RW, Grief GG, van Grondelle R, Hunter CN (1992) Nature 355: 848
150. Brunisholz RA, Zuber H (1993) Photochem. Photobiol. 57: 6
151. Breton J, Vermeglio A (1982) In: Govindjee (ed) Energy conversion by plants and bacteria. Academic, New York, p 153 (Cell Biology: A Series of Monographs. Photosynthesis vol 1)
152. Tapie P, Choquet Y, Breton J, Delepelaire P, Wollman F-A (1984) Biochim. Biophys. Acta 767: 57
153. Breton J, Katoh S (1987) Biochim. Biophys. Acta 892: 99
154. van Dorssen RJ, Breton J, Plijter JJ, Satoh K, van Gorkom HJ, Amesz J (1987) Biochim. Biophys. Acta 893: 267
155. Breton J, Duranton J, Satoh K (1988) In: Scheer H, Schneider S (eds) Photosynthetic light-harvesting systems-Organisation and function: proceedings of an international workshop, 12–16 October 1988. Freising, Germany. Walter de Gruyter, Berlin p 375
156. Hemelrijk PW, Kwa SLS, van Grondelle R, Dekker JP (1992) Biochim. Biophys. Acta 1098: 159
157. Kazachenko LP (1965) Opt. Spectrosc. 18: 397
158. Jennings RC, Zucchelli G, Garlaschi FM (1991) Biochim. Biophys. Acta 1060: 245
159. Kasha M (1963) Radiation Research 20: 55
160. Förster Th (1965) In: Sinanoglu O (ed) Delocalized excitation and excitation transfer. Academic, New York p 93 (Modern Quantum Chemistry, Part III B: Light and Organic Crystals)
161. Dexter DL (1953) J. Chem. Phys. 21: 836
162. Pearlstein RM (1982) In: Govindjee (ed) Energy conversion by plants and bacteria. Academic, New York, p 293 (Cell Biology: a Series of Monographs. Photosynthesis, vol 1)
163. Kenkre VM, Knox RS (1974) Phys. Rev. Letters 33: 803
164. Rahman TS, Knox RS, Kenkre VM (1979) Chem. Phys. 44: 197
165. Knox RS, Gülen D (1993) Photochem. Photobiol. 57: 40
166. Shipman LL (1977) Photochem. Photobiol. 26: 287
167. Sauer K, Lindsay-Smith JR, Schultz AJ (1966) J. Am. Chem. Soc. 88: 2681
168. Bittner Th, Voigt J, Irrgang K-D, Renger G (1993) Photochem. Photobiol. 57: 158
169. van Grondelle R, Amesz J (1986) In: Govindjee, Amesz J, Fork DC (eds) Light emission by plants and bacteria. Academic, New York, p 191 (Cell Biology: A Series of Monographs)
170. Knox RS (1975) In: Govindjee (ed) Bioenergetics of photosynthesis. Academic, New York, p 183 (Cell Biology: A Series of Monographs)
171. Jia Y, Jean JM, Werst MM, Chan C-K, Fleming GR (1992) Biophys. J. 63: 259
172. Chang JC (1977) J. Chem. Phys. 67: 3901
173. Avarmaa R, Kochnbey SM, Tamkivi RP (1979) FEBS Lett. 102: 139
174. Mimuro M (1988) In: Scheer H, Schneider S (eds) Photosynthetic light-harvesting systems-Organisation and function: proceedings of an international workshop, 12–16 October 1988. Freising, Germany. Walter de Gruyter, Berlin, p 589
175. Mimuro M, Jamai N, Yamazaki T, Yamazaki I (1987) FEBS Lett. 213: 119
176. Knox RS, Lin S (1988) In: Scheer H, Schneider S (eds) Photosynthetic light-harvesting systems-Organisation and function: proceedings of an international workshop, 12–16 October 1988. Freising, Germany. Walter de Gruyter, Berlin, p 567
177. Holzwarth AR, Haehnel W, Ratajczak R, Bittersmann E, Schatz GH (1990) In: Baltscheffsky M (ed) Current research in photosynthesis, Kluwer, Dordrecht, vol 2 p 611
178. Lin S, van Amerongen H, Struve WS (1992) Biochim. Biophys. Acta 1140: 6

179. Roelofs TA, Gilbert M, Shuvalov VA, Holzwarth AR (1991) Biochim. Biophys. Acta 1060: 237
180. Wasielewski MR, Johnson DG, Govindjee, Preston C, Seibert M (1989) Photosynth. Res. 22: 89
181. Durrant JR, Hastings G, Joseph DM, Barber J, Porter G, Klug DR (1992) Proc. Natl. Acad. Sci. USA 89: 11632
182. Roelofs TA, Kwa SLS, van Grondelle R, Dekker JP, Holzwarth AR (1993) Biochim. Biophys. Acta 1143: 147
183. Gillbro T, Sundstrom V, Sandstrom A, Sprangfort M, Andersson B (1985) FEBS Lett. 193: 267
184. Kwa SLS, van Amerongen H, Lin S, Dekker JP, van Grondelle R, Struve WS (1992) Biochim. Biophys. Acta 1102: 202
185. Eads DD, Castner EW, Alberte RS, Mets L, Fleming GR (1989) J. Phys. Chem. 93: 8271
186. Shipman LL, Housman DL (1979) Photochem. Photobiol. 29: 1163
187. Reddy NRS, Lyle PA, Small GJ (1992) Photosinthesis Res. 31: 167
188. Werst M, Jia Y, Mets L, Fleming GR (1992) Biophys. J. 61: 868
189. Owens TG, Webb SP, Alberte RS, Mets L, Fleming GR (1988) Biophys. J. 53: 733
190. Owens TG, Webb SP, Eads DD, Alberte RS, Mets L, Fleming GR (1987) In: Biggins J (ed) Progress in photosynthesis research, vol 1. Martinus Nijhoff, The Hague, p 83
191. Pearlstein RM (1982) Photochem. Photobiol. 35: 835
192. Gillbro T, Sandsrom A, Spangfort M, Sundstrom V, van Grondelle R (1988) Biochim. Biophys. Acta 934: 369
193. Paillotin G, Swenberg CE, Breton J, Geacintov NE (1979) Biophys. J. 25: 513
194. den Hollander WTF, Bakker JGC, van Grondelle R (1983) Biochim. Biophys. Acta 725: 492
195. Dainese P, Marquardt J, Pineau B, Bassi R (1992) In: Murata N (ed) Research in photosynthesis. Kluwer, Dordrecht, vol 1 p 287
196. Siefermann-Harms D (1987) Physiol. Plantarum 69: 561
197. Tanada T (1951) Am. J. Bot. 38: 267
198. Barret J, Anderson JM (1977) Plant Sci. Letters 9: 275
199. Trautman JK, Shreve AP, Owens TG, Albrecht AC (1990) Chem. Phys. Lett. 166: 369
200. Shreve AP, Trautman JK, Owens TG, Albrecht AC (1991) Chem. Phys. Lett. 178: 89
201. Wasielewski MR, Kispert LD (1986) Chem. Phys. Letters 128: 238
202. Gillbro T, Cogdell RJ (1989) Chem. Phys. Letters 158: 312
203. Bondarev SL, Bachilo SM, Dvornikov SS, Tikhomirov SA (1989) J. Photochem. Photobiol. A46: 315
204. Mimuro M, Nagashima U, Takaichi S, Nishimura Y, Yamazaki I, Katoh T (1992) Biochim. Biophys. Acta 1098: 271
205. Shreve AP, Trautman JK, Owens TG, Albrecht AC (1991) Chem. Phys. 154: 171
206. Wasielewski MR, Jhonson DG, Bradford EG, Kispert LD (1989) J. Chem. Phys. 91: 6691
207. Owens TG, Shreve AP, Albrecht AC (1992) In: Murata N (ed) Research in photosynthesis. Kluwer, Dordrecht, vol 1 p 179
208. Katoh T, Nagashima U, Mimuro M (1991) Photosynth. Res. 27: 221
209. Mimuro M, Nagashima U, Nagaoka S-I, Takaichi S, Yamazaki I, Nishimura Y, Katoh T (1993) Chem. Phys. Letters 204: 101
210. Razi Naqvi K (1980) Photochem. Photobiol. 31: 523
211. Cogdell RJ, Frank HA (1987) Biochim. Biophys. Acta 895: 63
212. Kohler BE, Spangler C, Westerfield C (1988) J. Chem. Phys. 89: 5422
213. Cosgrove SA, Guite MA, Burnell TB, Christensen RL (1990) J. Phys. Chem. 94: 8118
214. Ruban AV, Horton P (1992) Biochim. Biophys. Acta 1102: 30
215. Hashimoto H, Koyama Y (1989) Chem. Phys. Lett. 154: 321
216. Thrash RJ, Fang H, Leroi GE (1977) J. Chem. Phys. 67: 5930
217. Thrash RJ, Fang H, Leroi GE (1979) Photochem. Photobiol. 29: 1049
218. Hashimoto H, Koyama Y (1989) Chem. Phys. Lett. 163: 251
219. Seely GR (1973) J. Theor. Biol. 40: 173
220. Fetisova ZG, Borisov AYu, Fok MV (1985) J. Theor. Biol. 112: 41
221. Owens TG, Webb SP, Mets L, Alberte RS, Fleming GR (1987) Proc. Nat. Acad. Sci. USA 84: 1532
222. van Grondelle R, Bergstrom H, Sundstrom V, van Dorssen RJ, Vos M, Hunter CN (1987) In: Scheer H, Schneider S (eds) Photosynthetic light-harvesting systems-Organisation and function: proceedings of an international workshop, 12–16 October 1988. Freising, Germany. Walter de Gruyter, Berlin, p 519

223. Van Dorssen RJ, Hunter CN, van Grondelle R, Korenhof AH, Amesz J (1988) Biochim. Biophys. Acata 932: 179
224. Bergstrom H, van Grondelle R, Sundstrom V (1989) FEBS Lett. 250: 503
225. Sundstrom V, van Grondelle R (1990) J. Opt. Soc. Am. 7: 1595
226. Wittmerhaus BP (1987) In: Biggins J (ed) Progress in photosynthesis research vol 1, Martinus Nijhoff, The Hague p 75
227. Mimuro M (1992) In: Murata N (ed) Research in photosynthesis. Kluwer, Dordrecht, vol 1 p 259
228. Mukerji I, Sauer K (1993) Biochim. Biophys. Acta 1142: 311
229. Fischer MR, Hoff AJ (1992) Biophys. J. 63: 911
230. Holzwarth AR (1992) In: Murata N (ed) Research in Photosynthesis. Kluwer, Dordrecht, vol 1 p 187
231. Trissl H-W (1993) Photosynth. Res. 35: 247
232. Butler WL, Kitajima M (1975) Biochim. Biophys. Acta 396: 72
233. Butler WL, Strasser R (1977) Proc. Natl. Acad. Sci. USA 74: 3382
234. Schatz GH, Brock H, Holzwarth AR (1987) Proc. Natl. Acad. Sci. USA 84: 8414
235. Schatz GH, Brock H, Holzwarth AR (1988) Biophys. J. 54: 397
236. Leibl W, Breton J, Deprez, Trissl H-W (1989) Photosynth. Res. 22: 257
237. Montroll EW (1969) J. Math. Phys. 10: 753
238. Liu B, Napiwotzki A, Eckert H-J, Eichler HJ, Renger G (1993) Biochim. Biophys. Acta 1142: 129
239. Schlodder E, Brettel K (1988) Biochim. Biophys. Acta 933: 22
240. Schlodder E, Brettel K (1990) In: Baltscheffsky M (ed) Current research in photosynthesis. Kluwer, Dordrecht, vol 1 p 447
241. Takahashi Y, Hansson O, Mathis P, Satoh K (1987) Biochim. Biophys. Acta 893: 49
242. Hansson O, Duranton J, Mathis P (1988) Biochim. Biophys. Acta 932: 91
243. Turconi S, Schweitzer G, Holzwarth AR (1993) Photochem. Photobiol. 57: 113
244. Holzwarth Ar, (1986) Photochem. Photobiol. 43: 707
245. Hodges M, Moya I (1986) Biochim. Biophys. Acta 849: 193
246. Owens Mets L, Alberte RC, Fleming GR (1989) Biophys. J. 56: 95
247. Wasielewski MR, Johnson DG, Seibert M, Govindjee (1989) Proc. Natl. Acad. Sci. USA 86: 524
248. Demmig-Adams B, Adams WW (1992) Ann. Rev. Plant Physiol. Plant mol. Biol. 43: 599
249. Ruban AV, Rees D, Noctor GD, Young A, Horton P (1991) Biochim. Biophys. Acta 1059: 355
250. Jennings RC, Garlaschi F, Zucchelli G (1991) Photosynth. Res 27: 57
251. Dainese P, Santini C, Ghiretti-Magaldi A, Marquardt J, Tidu V, Mauro S, Bergantino E, Bassi R (1992) In: Murata N (ed) Research in photosynthesis, Kluwer, Dordrecht, vol 2 p. 13
252. Jansson S (1992) PhD Thesis, University of Umea, Sweden

Received: March 1994

Author Index Volumes 151-177

Author Index Vols. 26-50 see Vol. 50
Author Index Vols. 51-100 see Vol. 100
Author Index Vols. 101-150 see Vol. 150

The volume numbers are printed in italics

Springer-Verlag and the Environment

We at Springer-Verlag firmly believe that an international science publisher has a special obligation to the environment, and our corporate policies consistently reflect this conviction.

We also expect our business partners – paper mills, printers, packaging manufacturers, etc. – to commit themselves to using environmentally friendly materials and production processes.

The paper in this book is made from low- or no-chlorine pulp and is acid free, in conformance with international standards for paper permanency.

Printing: Saladruck, Berlin
Binding: Buchbinderei Lüderitz & Bauer, Berlin